HEARTIFICIAL INTELLIGENCE

HEARTIFICIAL INTELLIGENCE

Embracing Our Humanity to Maximize Machines

JOHN C. HAVENS

JEREMY P. TARCHER/PENGUIN
an imprint of Penguin Random House
New York

Jeremy P. Tarcher/Penguin
An imprint of Penguin Random House LLC
375 Hudson Street
New York, New York 10014

Library of Congress Cataloging-in-Publication Data

Names: Havens, John C., author.
Title: Heartificial intelligence : embracing our humanity
to maximize machines / by John C. Havens.
Description: New York : Jeremy P. Tarcher/Penguin, [2016] | Includes
bibliographical references and index.
Identifiers: LCCN 2015040084 | ISBN 9780399171710
Subjects: LCSH: Artificial intelligence—Social aspects. | Information technology—
Social aspects. | Human-computer interaction—Social aspects.
Classification: LCC Q335 .H3927 2016 | DDC 303.48/34—dc23 LC record
available at http://lccn.loc.gov/2015040084

Printed in the United States of America
10 9 8 7 6 5 4 3 2 1

Book design by Michelle McMillian

For Nate and Sophie,
my greatest contributions toward an authentic future

A QUICK NOTE REGARDING THIS BOOK'S FORMAT

Hello, and welcome to *Heartificial Intelligence*!

We have some exciting news! This book has been created using an old-fashioned publishing process utilizing paper and ink. Our historical research indicates this format allows humans to read, ruminate, and react to ideas without the need to click away to fourteen cat videos, Facebook posts, or tweets.* Our focus groups also indicate that this publishing format will help reinforce your sense of messy yet glorious humanity by forcing you to confront your own thoughts untainted by algorithmic influence.

Furthermore, outside of information regarding your initial purchase of this book, your actions will not be tracked in any way once you start reading it.** While it's tempting to try and influence your reaction to the book by modern tracking and profiling methodologies, the title of the book indicates our desire for you to take the time you deserve to analyze how emerging technologies are affecting your humanity.

Apparently humans are equipped with hearts and minds of their own.*** So our advice is to use the ones you already have to increase happiness and well-being before relying on the external ones other people are currently building. Not that these people aren't building amazing and worthwhile things, mind you. But our feeling is you won't be able to fully appreciate *artificial* intelligence until you define your own *genuine human values* first.

Thanks for your time. We hope you enjoy this more traditional process of reading and the personal introspection we've heard it provides.

You're worth it.****

* If you've opted to purchase this text as an e-book and prefer to click away to support cat videos, Facebook posts, or tweets, we recommend stating in a loud voice, "I am a HUMAN and will not be tracked!" This will serve as a centering process to remind yourself of your inherent humanity due to your ability to publicly act illogically and with great fervor. Please note, however, that you will still be tracked by hundreds of external data brokers, advertisers, and other organizations, any of whom may try to sell you sexual vitamin supplements. We have only tried about seven of these and cannot legally attest to their efficacy.

** At least not by the author and publisher. People may stare at you while you're reading in Starbucks or your kids may distract you during the precious seven minutes available to you to read during the day since, if you're like me, you fall dead asleep at some embarrassing time like nine thirty because you're exhausted from parenting all day along with everything else in your life, right?

*** Many doctors have said this. At least one of them looks like Socrates, so we're pretty confident this is true.

**** Seriously, you are. If you're like me, artificial intelligence does one of three things to you:

- Terrifies you because you think your toaster is going to kill you.
- Concerns you because your boss just gave your bonus to an algorithm.
- Mystifies you because your humanity is managed more by machines every day.

Don't wait until the Singularity comes and artificial intelligence takes over the world to believe me on this. Toasters are mean little buggers.

AUTHOR'S NOTE

The challenge in writing about an emerging technology such as artificial intelligence is that between the time you finish your manuscript and when your book is published, there's a strong possibility a new discovery has been made in the field about which you've written. So, in an effort to placate any future commenters on Amazon, Reddit, or any other platform:

- At the time of writing this book, Google has still not announced specifics around its artificial intelligence ethics board. If by the time you're reading this, it has, I say thank you to Google! Hopefully I'm expressing my gratitude while even being on that board, enjoying delicious and healthy meals from Google's famed cafeterias between stints in one of their nap pods (where I do my best thinking).
- At the time of writing this book, there is no formal, industry-wide AI ethical standard of which I am aware. I certainly could have missed it, and there are a number of great organizations working in this area, many of whom I interviewed in this book. My focus in *Heartificial Intelligence* is more to point out the necessity of such standards and why they're important to individuals than to comment on one organization's accuracy or efficacy.

- If we've been taken over by AI/robots in some way, then I'd like to point out how favorable I am regarding our future sentient colleagues. Seriously. This is not an anti-AI book, or anti-transhumanism book. I believe AI ethics is actually a more mature way of looking at our future with machines than creating potential sentient machines we sell and enslave. Call me crazy.
- I think it's important to have a sense of humor about AI. This doesn't mean I take the subject lightly. As I point out in the book, I have struggled a great deal with fear and depression on the subjects of automation and loss of agency. But I wrote this book to help you move beyond fear and be constructive as you consider the inevitability of AI in our midst, today. My vote is that we greet the future with curiosity, laughter, and joy versus dread, gloom, and fear.

FINAL AUTHOR'S NOTE

I'm a huge fan of Monty Python, so this author's note serves no purpose except to be silly.

CONTENTS

INTRODUCTION

Spring 2021

"If you want your daughter to live, this is the only solution."

My wife was in the waiting room with my two kids, my eleven-year-old son and my nine-year-old daughter, the white paper on the examining table freshly crinkled from where Melanie had been examined moments before. The smell of the alcohol swab they'd used after taking her blood still hung in the air.

"So the computer chip goes directly in her brain?" I asked again. I was having a difficult time understanding what exactly was going to happen to my daughter to combat her young-onset Parkinson's disease.[1] A year before, her hands had begun shaking throughout the day. Her seizures increased in intensity, and two months ago she began experiencing blackouts and fell down at school. Her diagnosis came quickly, although she'd gone through a battery of painful tests to confirm it was Parkinson's.

"Yes," answered Dr. Schwarma, our family practitioner for the past six years. An extremely sharp and caring woman in her midthirties, she never beat around the bush with her diagnoses. She'd contacted a friend who worked in Manhattan who specialized in the procedure. "The chip will help control the erratic synapses in her brain that are causing her seizures."

I pointed to the iPad in her hands. "Is the chip like something you'd find in a computer? I'm assuming it stays in her brain permanently once it's put in?"

"That's the hope, although the human body is an intense environment. There's a good chance the chip will need to be replaced, but it's a relatively simple procedure even though it involves the brain. Plus, there's the possibility of remote updates for the chip with newer technology, which would mean less chance of future surgery."

I paused before speaking as the voice of the office secretary came over a loudspeaker calling for one of Dr. Schwarma's colleagues to come to the front desk. "So if there are remote updates," I said, "this chip will be firmware, correct? It's not, like, the silicon equivalent of a stent or whatever; it's active technology."

Dr. Schwarma nodded. "That is correct."

"So that will involve Wi-Fi or Bluetooth or iBeacon technology or whatever."

She nodded again. "I'm not sure about the specifics, but the basic logic is that we'll need to remotely check on the status of the chip's operation without performing surgery. So some short-range technology like one you've mentioned will be used."

"So she could be hacked?" My chest got tight and I felt my eyes moisten. "Right? And how does the Wi-Fi stuff work? Does she have a passcode for her brain? And can she travel? How does she explain this to the TSA in airports?"

Dr. Schwarma held up her hand. "John—those are all important questions and there will certainly be challenges ahead. But the positives far outweigh the negatives."

"I'm sorry," I answered as I wiped my eyes with the back of my hand. "It's just freaky to picture a chip in my daughter's brain. Could she eventually update the chip to be an internal smartphone? Be her own Wi-Fi hot spot? And does this make her a cyborg?"

Dr. Schwarma shook her head. "Cyborg choices typically involve a person replacing parts of their body outside of a life-threatening need. However, technically she will be part machine." She held up her mobile phone. "No more than the rest of us, of course."

"But we can turn our phones off," I answered. "The chip will always be with her."

She took a step forward, laying her hand on my shoulder. "Yes, the chip will always be with your daughter, John. But unless you do this procedure, she won't be."

The Genuine Challenge

A few years back I wrote an article about artificial intelligence (AI) for Mashable, a popular online news site focused on technology and culture. My goal was to evolve the conversation around AI beyond the polarized views of complete acceptance and rejection of the technology. While I believe AI is inevitable in our lives, I don't believe that means we should blindly accept whatever new development in the field comes down the pike. Likewise, living in fear about the evolution of the technology doesn't help humanity either. For my article, I really wanted to identify some potential solutions regarding humans working or joining with machines that I could wrap my brain around.

Initially, my research depressed me a great deal. I learned how quickly the AI field is growing without there being industry-wide standards around safety for development. I learned nobody has clarity regarding if and when machines might become sentient (intelligent and "alive"), but multiple experts who said that could never happen had been recently surprised at advances that were changing their minds. Overall, I've come to learn that whether or not machines become truly sentient, the widespread adoption of AI is inevitable. And while people developing or utilizing AI keep saying, "We need to make sure we understand the ethical issues around this technology," they nonetheless keep building systems they may not be able to control.

I see this as a problem.

And an opportunity.

My Mashable article expanded to become this book, and what I came to realize after years of research and interviews is there are no simple answers regarding the evolution of AI. Nobody can accurately predict when machines or robots will "come alive" or exactly how that will look.

So for my part, as an exercise to deal with my concerns, I began to imagine personal scenarios in which I couldn't avoid AI in my life. That's how I came to the fictional scenario about my daughter you just read. As much as I may fear aspects of AI, if a piece of technology would mean the difference between my daughter (who is real) living or dying, I'd utilize the technology.

While imagining along these lines may seem strange, the process provided catharsis for me. Instead of being anxious about a future dominated by machines, I began to more deeply examine issues of AI as inspiration to validate my humanity. That's why every chapter of this book opens with a fictional vignette—I want to help you move beyond the polarizing debate around AI and imagine how you'd react to the scenarios I present. AI is not just science fiction any longer. It's here. My goal with these stories is to help you more rapidly go through the journey I did of genuinely confronting my fears to get to a positive place regarding the inevitability of AI. The body of each chapter describes the tech and issues I bring out in the fictional vignettes.

I do have a warning for you, but it's not about killer robots taking over the world within a few decades. The field of AI is advancing so rapidly we may lose the opportunity for introspection unhindered by algorithmic influence within a few *years*. Many of us are already at the point where we look to our devices and the code that drives them to make every major decision in our lives: Where should I go? Whom should I date? How do I feel? These "digital assistants" are hugely helpful tools.

But they've also trained us to delegate decisions as a default. This process involves a willingness to sacrifice the parts of ourselves that used to make these decisions to technology. For my part I can live without my kids ever knowing how to use a paper map, but I'm not comfortable with their potential inability to identify a life partner without the aid of an algorithm. I can live with apps that monitor my heartbeat and brain waves to help me identify when I'm happy. I'm not comfortable with devices that manipulate these insights to motivate behavior I don't fully understand.

Technology has been capable of helping us with tasks since humanity began. But as a race we've never faced the strong possibility that machines may become smarter than we are or be imbued with consciousness. This technological pinnacle is an important distinction to recognize, both to elevate the quest to honor humanity and to best define how AI can evolve it.

That's why we need to be aware of which tasks we want to train machines to do in an informed manner. This involves individual as well as societal choice. We're at a tipping point in human history, where delegating as a habit may lead us to outsource aspects of our lives we'd benefit more from experiencing ourselves. But how will machines know what we value if we don't know ourselves?

That's the genuine challenge, and the basis for *Heartificial Intelligence*—on an individual level, and for humanity as a whole. That's also why the subtitle for the book is *Embracing Our Humanity to Maximize Machines*. We need to codify our own values *first* to best program how artificial assistants, companions, and algorithms will help us in the future.

This concept is your genuine challenge as well, and why I've written this book.

And to be clear if you're a geek like me and think I'm dissing technology: I am *not* anti-AI. I'm pro-human. These are not mutually exclusive. If machines are the natural evolution of humanity, we owe it to ourselves to take a full measure of who we are right now so we can program these machines with the ethics and values we hold dear. In AI, there's a concept known as deep learning[2] that describes an approach[3] to building neural networks based on machines learning methods of observation. My recommendation is that we apply a similar deep learning process for our own lives based on codifying the ethics, values, and attributes unique to humanity.

Some good news: There's a science known as positive psychology that's helping individuals increase their well-being after observing how actions such as gratitude and altruism have improved their lives. I'm using the term *well-being* as it refers to the intrinsic, long-term increase in life satisfaction these actions can bring versus a fleeting, mood-based happiness. While this "hedonic happiness" is natural and lovely, positive psychology has shown that constantly trying to improve your mood is both erratic and exhausting. Genuine flourishing, a holistic state involving your mental, physical, and spiritual well-being, is achieved by repeating actions that provide insights not based solely on emotion. This is a form of deep learning we should apply to our lives.

Some challenging news, however: You can't automate your well-being. While you can utilize an app to keep a gratitude journal or measure your blood pressure during meditation, a machine *can't experience your well-being for you*. Not yet, anyway. This is not meant to be pejorative toward the potential of AI or machines but to simply acknowledge they're built differently than people. *Automated* happiness doesn't work for humans, according to positive psychology. Delegating core emotional or spiritual work doesn't compute. Predictive algorithms can help provide insights that affect our mood but the increase of long-term well-being involves our conscious and ongoing involvement.

A bit of hard truth here that needs acknowledging: In many ways, it's actually easier to delegate decisions around our well-being to machines or to avoid deeper questions about what makes us happy or human. But this book is not a formulaic, "get happy quick" scheme to deal with the inevitability of a dark AI future. It's about testing solutions that validate you're worth a deeper look.

A Vision for Values

While positive psychology is having a transformational effect on people around the world, it can't improve our lives if we're discouraged from looking within. Here's why:

- More than automation removing our jobs, preferential algorithms may remove our desire for introspection.
- More than *Terminator* robots killing us with guns, artificial intelligence may replace our ability to think on our own.
- More than confusion over how our consciousness could be replicated in machines, we are currently prioritizing the opportunities these technologies *may* bring over defining the benefits humans *already possess.*

It's this third point that's the inspiration for this book. In terms of automation, comparisons between machines and humans typically revolve around questions of skill. This is a lamentable irony when you consider we've built AI systems specifically to replicate our tasks in the first place. At best, it's a temporary comfort to wonder which skills machines may possess or when.

What humans currently have that machines do not, however, is an inherent sense of values. We develop these over time based on our environment, but we're also equipped with an emotional and moral sensibility that machines don't currently share. While advances in fields like cognitive computing may evolve to the point that companion robots appear to have emotions, their ethical behavior will initially be based on the humans who programmed them. This is why, in a very real sense, the future of our happiness is dependent on teaching machines what we value the most.

And I mean this literally. I believe as individuals and as a society we need to identify, track, and codify our values so we can translate them into machine-readable protocols. It's okay if you think that sounds crazy difficult. It is. But so is trying to create sentient machines. And ironically enough, a lot of AI methodologies revolve around observing our ethical behavior as demonstrated by our actions. So they're already codifying our values, oftentimes without our direct input. This means lethal autonomous weapons (machines that can kill without direct human intervention) will act based on whatever country's programmers created them. Or your self-driving car may be programmed to hit an errant pedestrian versus risk hurting you based on decisions made by the car's manufacturer.

How do you feel about that? Should *your* values or ethics inform these decision-based protocols?

Yes, they should. Otherwise, your values will be ignored in the sense that all devices and products will favor the ethical biases of the programmers who created them. That doesn't mean they're bad people—they're just not you. What if your faith dictated in an accident involving a self-driving car that you would want to give your own life to spare someone else? Why shouldn't the car or product you've purchased reflect this desire? Jason Millar, a philosophy professor at Queen's University in Kingston, Canada, calls this concept "technology as moral proxy,"[4] which provides a huge opportunity for innovation versus just regulation. Like the precedent of informed medical consent, a codified ethical framework for humans living with AI would provide legal clarification around situations we're going to be facing a lot in the near future. It would also broadcast personalization data based on your values that would allow companies and individuals to be deeply sensitive to your needs.

I call this codification of our ethical choices Values by Design, and in the latter part of this book I've provided a framework for you to track and codify your values based on established psychological research. It's a pretty simple process: There are twelve core values (family, health, etc.) that you rank on a scale of 1 to 10. This provides a sort of ethical snapshot of your life, allowing you to clarify what values you hold most dear. Then for three weeks, at the end of every day you rank each of the twelve areas based on whether or not you lived to those values that day. So, for instance, say you value family as a 10 when you start your tracking. Then after three weeks, you realize you're not spending any time with your family (meaning you're ranking family at a low level every day). This insight will help you see where your life may be out of balance, and how you can adjust your actions based on the data reflecting how you actually live your life.

It's a simple process on purpose. It can be enhanced with apps monitoring your heart rate or stress levels, but a core part of its benefit comes from daily reflection on how you've lived your life.

And it's amazing how few people I talk to can even *name* five top values they pursue every day. Even fewer have ever tested them in any meaningful way. Of course religion, faith, and other methodologies focused on values have helped us refine our ethical decisions over the years. But my goal with Values by Design is to present a framework for this tracking process that could potentially complement AI systems and data measuring us in the same way. In that sense, everyone involved will know we've taken the time to substantiate the values we most want to reflect.

P.S., I'm not arrogant enough to think Values by Design is *the* process to save the world by providing an ethical solution for our adoption of Artificial Intelligence. I'm simply championing one way for an individual to track his or her values that could also inform the morally oriented decisions being made by machines.

This is why I've also dedicated a great deal of this book to highlighting the field of ethics in artificial intelligence, as I believe it provides the key to moving forward effectively with humans and machines. I believe ethical programming has to be imbued at the manufacturing stage of any AI system to ensure it's safe, useful, and relevant for society at large.

This means we have every reason to allow ourselves to identify what we value most and to live our lives in accordance with those ideals. In fact, it's a mandate that we all undergo this process, or machines will base their ethical programming on examples provided by YouTube or *The Real Housewives of New Jersey.*

It's a deep challenge to name and track the specific values we live according to every day. But the process allows us to see where we're out of balance regarding money, time, health, or any other metrics providing meaning in our lives. Taking the time to measure these things is what brings authentic purpose to our lives.

It's what makes us genuine.

The Deal About Our Data

While most people would hardly consider a chip to monitor brain activity something that could transform a person into a robot, the fictional scenario about my daughter provides a physical example of the communion we already share with machines. Our computers, mobile phones, and the objects around us connect us to the Internet—and, subsequently, to data that we input into our minds and hearts on an almost constant basis. In return, our thoughts and actions create data that enters the vast pool of information swirling around us at all times, unseen yet very real all the same.

Google Glass introduced the general population to augmented reality (AR), technology that overlays digital information about your surroundings onto the lens through which you see the world. Oculus Rift, acquired by Facebook, is a highly advanced form of virtual reality (VR), in which your eyes and ears are covered while you're immersed in a video game or other sensorial experience. Whatever the interface, all the hardware simply provides intermediary steps for us to get used to the inevitable union of humans and machines—or more specifically, the *physical* union of humans and machines. As Dr. Schwarma pointed out in my vignette, the mental and behavioral union has already taken place.

The physical issues are relatively simple. Today people are excited about wearable technology, in which they've traded the design and user interface of a mobile phone for a piece of clothing or jewelry like the Apple Watch. Soon, augmented reality contact lenses will replace mobile phones altogether, with some people opting for LASIK surgery so the technology need never be removed. We've become accustomed to the idea of technology-enhanced prostheses for athletes,[5] like sprint runner Oscar Pistorius's controversial prostheses, which earned him the nickname the Blade Runner. Now it's just a matter of personal choice as to when to marry carbon with silicon.

But the fact that our lives are represented by our personal data in the digital realm is still a relatively new concept to most people. We understand that we have different personae in various digital arenas—we act more professionally on LinkedIn, more laid-back on Facebook. But this all involves data *we can see* and that we knowingly create. But our holistic and hidden digital identity is defined by the actions we take online and off that are tracked at all times. And the overwhelming majority of organizations doing the tracking don't share the insights they glean about our lives with us.

Machines in this context are utilized to create algorithms that can best analyze and predict our future behaviors. In a very real sense, organizations with access to our aggregate identity know more about us than we know about ourselves. Edward Snowden helped turn the tide on this lack of knowledge regarding governmental surveillance of our lives. But while state-driven tracking issues are certainly critical to consider, they're not the focus of this book.

Why It's *Artificial* Intelligence

They say you can't stop progress. But we can redefine it.

Like most people, my first exposure to artificial intelligence came from science fiction movies such as *The Terminator*. It's easy to get caught up in the idea of robots getting smarter than us and destroying the human race. However, it's stories like *Minority Report* that I find much more intriguing and ominous. Tom Cruise's character in the movie is head of a futuristic police division known as PreCrime, working with cyborg "precogs" (clairvoyant humans synced with machines) to help identify people who are intending to break the law. The Internet economy is currently driven by technologies working to identify our intentions toward purchase in a similar way, predicting and controlling the outcome of our actions.

The sinister aspect of this tracking is not about commerce. It's about the lack of transparency regarding our data that's at the heart of the system controlling that commerce. Internet and mobile advertising are built on surveillance, tracking our behavior to see when we're most likely to purchase. We're called "consumers" in this model because that's our primary role to play—to buy items to provide further insights about what else we'll buy. While companies are trying to improve our lives with products or services that consumers genuinely need, purchase funnels never end in abstinence. Our actions are tracked so predictive algorithms can analyze our behaviors to generate messages that will inspire further purchase.

This is why this is *artificial* intelligence—as humans we're built for purpose, not just purchase.

Knowing about ourselves only in the context of purchase provides a shallow picture of our whole persona. In a world where gross domestic product (GDP) is our primary measure of value, we've been led to think greater productivity or profits are the keys to human happiness. If this were true, tracking and manipulating purchase behavior to increase people's happiness would make a great deal of sense. If buying certain products or spending money just for the sake of it made us happier, our lives would be a lot simpler (if we had the cash to support this hypothesis). But the science of positive psychology has demonstrated that intrinsic well-being, or "flourishing," is not increased by a surplus of money. While we need a basic amount of material goods to feel safe and secure, intrinsic well-being is increased by actions such as mindfulness or altruism, expressing gratitude or doing work that brings us "flow." In the same way we can go to the gym to increase our physical well-being, we can repeat actions that will increase our happiness. It's not a formula, however. It's a journey.

The challenge for artificial intelligence in this context is determining where learning algorithms improve a person's life versus cutting corners on his conscious efforts to improve his well-being. For instance, I'll trade the loss of serendipity when searching for a book on Amazon, knowing I won't ever be shown a horror novel. But advertisements appearing in my Facebook feed touting products that will supposedly increase my happiness are shrouded in alarming mystery. What behaviors of mine have been tracked to result in my seeing this ad? If you have insights about my well-being, why won't you provide them to me? I'd gladly buy a product from someone willing to share these precious details. But Google and Facebook rely on the hidden nature of data collection to sustain their business models of advertising. It's not in their best interests to share data about our unique identities and actions. But regarding our happiness, how can we accurately measure which actions improve our well-being when companies won't reveal insights based on our lives?

This is also why it's *artificial* intelligence—individuals don't currently control the data relating to their identity. IP trumps I.D.

The world of data and identity I've described may soon comprise the majority of our lives as revealed by the devices we wear. Today, we turn off our computers or put our phones aside, however briefly, and experience the world as seen through our own eyes. But once we're wearing the lenses, or browsers, at all times that reveal the invisible data surrounding our identity, we're in for a hell of a shock. In the fictional scenario I've described with Dr. Schwarma I was presented with a choice regarding the possibility of my daughter's visceral union with technology. As things stand now, we've passively accepted the loss of our personal data that fuels algorithms, AI, and the Internet economy as status quo. By the time we see how our identities look in the virtual world as controlled by others, it will be too late to get back the rights we've let go. It's time we stopped relying on artificial measures to increase our genuine well-being.

Happinomics

It wasn't until about three years ago that I realized economics is a study of philosophy as much as of statistics. Measuring and attributing value to individuals, communities, or countries requires universal agreement on which metrics to utilize before creating any kind of standard report regarding policy or welfare.

It's hard to imagine a time when GDP wasn't utilized as a measure for every country of the world to determine their well-being. The logic regarding the GDP is that as it goes up, a country's happiness increases as well. But this connection hasn't proved true, as the economist Simon Kuznets, who created the concept of GDP in the 1930s, predicted. As Lauren Davidson points out in her November 2014 *Telegraph* article "Why Relying on GDP Will Destroy the World," Kuznets warned, "The welfare of a nation can . . . scarcely be inferred from a measurement of national income such as GDP."[6] We didn't heed his warning, and the GDP was adopted as a set of standardized values everyone in the world agreed were the most important to measure.

Unfortunately, these values focused largely on metrics regarding income and growth while ignoring other issues of well-being and social justice. The values made their way to the business world, where increased profits and shareholder gains reflected and informed the GDP, shaping ideals regarding employee productivity and worth. Eventually these values made their way to the individual level. We're told it's our civic duty to consume goods to increase the economy, while also believing that having money and success leads to happiness. But this model hasn't panned out. Increasing GDP for a country could come in the form of removing oil from its region and destroying the environment. This means short-term gains adversely affect long-term sustainability. Likewise for individuals, getting a higher salary without a sense of purpose for your work doesn't equate to increased happiness.

Business owners also have to come to grips with the ultimate end of GDP regarding issues of automation and machine learning. If machines continue to excel in their ability to do our jobs, it will always be cheaper to utilize them in place of human beings. Note I didn't say "better"—but machines work without complaining, without the need for insurance, and without the need for a break. So in effect, organizations not questioning the status quo of a GDP-mandated focus on growth are implicitly inviting widespread automation of jobs by machines. This is a moral and ethical as well as fiscal decision for any organization to determine as part of their values. Do we want a human or a machine workforce? And is it realistic to think we can work alongside machines in a sustainable manner as they continue to learn and excel at the tasks we're teaching them?

In terms of the Internet economy, as long as a GDP-focused model of increased profits for shareholders continues to hold sway, companies such as Google and Facebook will continue to rely on clandestine, tracking-based advertising models for their revenue. Utilizing artificial intelligence in this context makes a great deal of sense since individuals are producing more personal data than ever before. Objects comprising the Internet of Things, like the Nest thermostat (owned by Google), will track ever more intimate aspects of our lives until we have no control over our data along with the thoughts, emotions, and behaviors that create it.

Some disclosure here: I'm not an economist by profession. But I have been evangelizing the adoption of new metrics beyond GDP such as gross national happiness and the genuine progress indicator for years. While this hardly qualifies me to be an economist, I am in a unique position with my background in technology to analyze how quantified data from an individual could affect policy creation at scale. I've also interviewed hundreds of experts in technology, economics, and positive psychology to determine what solutions could be pursued to deal with the artifice that's eroding trust in technocratic and GDP-driven environments today.

In my work as a technology writer (I'm a contributing journalist to Mashable and the *Guardian*), it's also become apparent we need to consider economic models that straddle the real and virtual worlds to employ metrics that can genuinely increase people's well-being. While it may seem strange to consider the economic dynamics within a MMORPG (massively multiplayer online role-playing game) affecting real-world markets, the rise of virtual currency and the amount of time people spend within these games is increasing exponentially and demands workable solutions.

When devices like Oculus Rift become ubiquitous, in which individuals cover their eyes and ears to become immersed in a virtual world, many people may choose to never again spend time in meatspace (a term that geeks like me use for the "real world"). Think how these dynamics will affect the GDP if it doesn't evolve to embrace the virtual realm.

What if someone has a job in-game that pays in Bitcoin? If their physical body resides in Dublin, Ohio, but the game's servers are in Dublin, Ireland, should the person pay U.S. or Irish taxes—or both? And of course, Facebook owns Oculus Rift, so we can assume Zuckerberg will utilize eye tracking, facial recognition, and stress-, heart-rate-, and brain-wave-sensing technologies to know how people are feeling within any game environment to provide real-time advertising opportunities to his clients. How will *that* type of guaranteed revenue model affect economics, let alone our mental psyche?

Genuine by Design

Heartificial Intelligence is focused on helping remove the artificial intelligence we've been exposed to in multiple arenas of our lives in exchange for new models of progress that can authentically improve well-being. It's designed to help you make informed, conscious choices regarding the technologies and values that will help you live an examined life.

I've broken the book into two sections: Artificial Intelligence and Genuine Progress.

The Artificial Intelligence section presents what I see as our current dystopian trajectory regarding well-being. While I see multiple positive aspects of AI, without the rapid influx of transparency and standardized ethics regarding its creation, it will continue to be dominated by technocratic ideals that mirror the values of GDP.

The Genuine Progress section presents multiple solutions to the issues I raise regarding artificial intelligence. These include technological, ethical, and economic examples in which I've worked to provide pragmatic solutions to issues described wherever possible. My hope is that by infusing transparency into the existing models and markets driving the world today we can benefit from the amazing world of AI versus being subsumed by it.

Here's a breakdown of the sections and chapters of the book. These are written as teasers rather than spoilers, to give you a specific sense of the issues and solutions the book describes:

SECTION ONE: ARTIFICIAL INTELLIGENCE

Chapter One: A Brief Stay in the Uncanny Valley

One of the chief ways many people are freaked out by robots is when they look too human. Remember the movie *The Polar Express*? While the animation was extremely advanced for its time, a lot of people were turned off when one of the characters exhibited traits that were almost but not quite human within their cartoon form. This concept is known as the uncanny valley in robotics. It's a widely accepted term for engineers, although some dislike its assumption that everyone will react the same way when seeing an android or robot for themselves. The concept is also mirrored within the realms of Internet tracking and advertising today. We all know our personal data is being tracked; we're just not sure how. Why do we keep seeing those diaper ads when we don't even have kids? Why does that same ad for a razor appear *every day* in my Facebook feed? While sentience and the Singularity are issues reflecting the future of AI, the algorithmic groundwork creating their future exists today and already deeply influences our digital identity.

Chapter Two: The Road to Redundancy

A study done by the University of Oxford, reported in 2014,[7] says, "Occupations employing almost half of today's U.S. workers, ranging from loan officers to cab drivers and real estate agents, [will become] possible to automate in the next decade or two." Similar statistics apply to the United Kingdom, as the *Telegraph* reported in a November 2014 article: "Ten million British jobs could be taken over by computers and robots over the next twenty years, wiping out more than one in three roles."[8] Automation by machines is a real threat to human employment and well-being in the very near future. While technological innovation and AI may bring great benefits to humanity, it will also change how we find meaning in our lives if we can't work. Beyond issues of pursuing purpose, we'll also need to be able to pay our bills in the wake of a machine-driven world.

Chapter Three: The Deception Connection

In a very real sense, artificial intelligence is focused on tricking us into thinking something is real that's not. It's a form of digital magic known as anthropomorphism, in which we might forget that Siri is a digital assistant and begin joking with it as if it's a person. While robot assistants for the elderly or Furbies for kids provide a great deal of comfort for those in need, they shouldn't be utilized *instead of* but as *complements to* human companionship.

Chapter Four: Mythed Opportunities

The AI driving advertising-based algorithms poses an existential threat to our well-being. When we're tracked solely or largely as a means of identifying what items we want to buy, the fuller context of who we are as human beings is lost. We begin wondering why we were offered a dietary supplement in response to a certain word we typed on Facebook—does some algorithm think I'm fat? Ethical standards for sociological study have always been the norm, in which participants are fully aware of what's being studied about their behavior. These guidelines need to be imbued within the economies comprising the Internet and Internet of Things to avoid an inevitable world of digital doppelgängers who only reflect our consumerist selves.

Chapter Five: Ethics of Epic Proportions

Robots don't have morals. They're physical objects imbued with code programmers have provided to seek established objectives. A big reason ethics is so critical to the AI industry is the lack of standard application for products today, especially at the design level of production. This is especially important regarding militarized AI, which has grown so rapidly in the past decade and received billions of dollars in funding. For civilians, issues of autonomous cars bring the need for AI ethics much closer to home. For instance, should a car entering a tunnel swerve to miss a child who has run in front of it to save her life even if it means killing the driver? This is a concept developed by Patrick Lin,[9] director of the Ethics and Emerging Sciences Group, based at California Polytechnic State University. Would you rather an expert in ethics like Patrick answer this question or a sleep-deprived programmer trying to make a deadline for an investor? We are in a unique era for issues and legal questions like these, as all law to this point has been written exclusively for humans. Robots are changing the rules, and policy needs to catch up to include their growing influence.

Chapter Six: Bullying Beliefs

You don't have to sit idly by to watch the future of humanity evolve without your input. Scientific determinism is as much of a faith as any major religion, if and when it proselytizes without permission or shifts cultural perceptions in dangerous directions. While we can't stop progress, questioning innovation based largely on creation of profit or growth is a necessity. It's not Ludditism to desire to move toward the future fully cognizant of what makes humanity glorious and laudable, as messy as it may be.

SECTION TWO: GENUINE PROGRESS

Chapter Seven: A Data in the Life

Privacy isn't dead; it's just been mismanaged. While a person's decision regarding her identity is her own business, it doesn't make sense to avoid creating frameworks for exchange of personal data that allow transactions to happen as a person desires. Whether it's personal clouds, vendor relationship management (VRM), or life management systems, there are multiple methodologies available today that will allow all people to protect and control their data in whatever ways they see fit. This will allow for greater accuracy regarding our data for any of the algorithmic or AI-oriented programs existing today or coming down the pike.

Chapter Eight: A Vision for Values **(How-to Chapter)**

To be genuine, you have to be able to articulate what you believe and prove to yourself you're living according to your values. In a future where our digital actions will be easily visible to other people via augmented or virtual reality, being accountable for our actions will be more important than ever. This chapter provides a step-by-step guide to identify and track your values based on the research of a number of respected experts in the field of sociology and positive psychology. By taking a measure of your life based on what you hold most dear, you'll be able to see what areas you may want to focus on more or less to achieve balance in your life. By identifying your values and taking actions based on your beliefs, you'll also be able to discover opportunities to help others in ways that will elevate your personal well-being.

Chapter Nine: Mandating Morals

Ethics in AI has never been more important. Recent announcements in the field, especially the Future of Life Institute's "Research Priorities for Robust and Beneficial Artificial Intelligence,"[10] provide encouraging precedents for experts to incorporate ethical guidelines in the core of all their work on AI and autonomous machines. But silos in academia or the bias of profit-first businesses cannot supplant the imbuing of human values into the core of our machines. We won't get a second shot to teach our successors right from wrong, so we need to understand and foster that process right away.

Chapter Ten: Mind the GAP (Gratitude, Altruism, and Purpose) **(How-to Chapter)**

In the UK when you take the Tube (or subway), you hear a pleasant voice warning you to "mind the gap" to keep from falling between the platform and the train. As a way to pursue *Heartificial Intelligence*, the science of positive psychology demonstrates how gratitude, altruism, and purpose can increase our intrinsic well-being. As compared to happiness, this form of flourishing is focused not on mood but on actions we take based on attributes that define who we are. This chapter contains background and exercises on how to perform a personal "GAP analysis" while also exploring why pursuing your purpose will become essential in a world where automation replaces our jobs. Whether we have governmental interventions to help pay our bills with a Basic Income Guarantee or other economic measures, minding the GAP will equip you to face the future knowing you can spend every day helping others and get happy in the process.

Chapter Eleven: The Evolution of Economics

You may have heard of gross national happiness, but you may not realize it's not focused on mood. Rather, it's a metric to help measure citizen well-being beyond the fiscal measures of the GDP. Newer metrics such as the Genuine Progress Indicator have also been adopted in states such as Maryland and Vermont and take the key business tenet of double-entry bookkeeping into account when trying to determine citizen well-being. Economically speaking, this means it factors in things such as environmental effects when an oil spill increases a country's GDP (since it creates jobs to clean up the mess). As sensors, wearable devices, and the Internet of Things more intimately measure our emotional and mental well-being, governmental metrics cannot simply focus on the increase of money or growth as an approximation of our happiness. This is also the chapter where I expand the concept of Values by Design, extending the pragmatic exercises I provide earlier in the book with a sense of how they can be incorporated into a digital and economic future driven by AI.

Chapter Twelve: Our Genuine Challenge **(Interactive Chapter)**

Remember the Choose Your Own Adventure series of books, in which you get to decide how a story unfolds? At the beginning of Chapter Twelve I've provided you with your own chance to do the same regarding Artificial Intelligence. This is also the chapter where I wrap up the idea of ethics needing to be incorporated into AI design and values into your life today.

From the Artificial to the Authentic

Tired of being freaked out about Artificial Intelligence?

Want to test what you value most in life?

Want to learn how positive psychology can improve your well-being?

Here's an invitation for you to answer these questions for yourself.

SECTION ONE
ARTIFICIAL INTELLIGENCE

one | A Brief Stay in the Uncanny Valley

Fall 2028

"Can I get you something to drink, Rob?" I asked, yearning for a stiff drink of my own.

"No thank you, sir, I'm good."

I pictured this moment hundreds of times over the past sixteen years, as I assume every dad does. My daughter Melanie's first date—or at least the first one where the guy came over to pick her up. Robert was a polite and good-looking guy with an athletic build, light brown skin, and piercingly blue eyes. In our first few minutes chatting while Melanie got ready upstairs with my wife, Barbara, Robert was a charming conversationalist. He seemed genuinely interested in my responses to his questions and was clever and funny without being snarky. I could see why Melanie was attracted to him, which is why I was working as hard as I could to keep from throwing up in my mouth.

Rob was a robot.

I only knew this because Melanie had told me the night before as preparation for Rob's visit. Barbara and I had kept asking questions about her mystery guy until she announced he was coming over.

Oh, and that he was a robot. Or as she put it, an "autonomous intelligence embodied in flesh form."

Humanistic robots by 2028 had become extremely advanced in terms of their physical appearance. They lost the "uncanny valley" effect of having just enough false movements or characteristics to remind people they weren't human. In the wave of automation that had begun around fifteen years ago, corporate executives had also begun replacing human workers with robots where it was deemed the machines could gain unique marketing insights from the people they served.

Initially most of the corporate robots looked like Baymax, the lovable inflatable helper-bot from the movie *Big Hero 6*. Then after people got used to anthropomorphizing robot workers in their cuddly manifestations, companies manufactured them in branded humanistic forms that best matched demographics of customers' nearby physical community. In Hendersonville, Tennessee, robot workers at Waffle House restaurants looked like Carrie Underwood. In Brooklyn Starbucks locations, Bruno Mars appeared to be the "botista" serving your lattes. Within a few years robot design improved to such an extent it wasn't necessary to have only a few male and female versions spread throughout a region. Company algorithms linking robots, people, and the connected objects around them predicted what type of interface a shopper wanted to deal with in public, and that was the literal face they'd see on their robot while conducting a purchase.

A majority of my friends lost their jobs in the wake of widespread automation. I'd only held on to my position as a technology writer by declaring myself a "human journalist." The idea started as a joke but when human writers were replaced by AI programs[1] management felt having a human reporter provided a form of objectivity my silicon counterparts lacked when reporting on technology.

Up until now, people's physical interaction with robots had been largely within the context of shopping or food service situations like at Starbucks. Of course pornbots had been going strong for years and I'd been on a few business trips where progressive hotels had begun listing robotic "services" as part of their adult offerings.

But this robot was dating my daughter. This was my Melanie. I realized that Rob was a far cry from an iPhone or smart fridge in terms of his advanced technology, but I couldn't stop feeling a sense of revulsion in his presence. As hard as I tried to be civil, I knew he was equipped with facial recognition and physiological sensor tracking and could tell I was freaking out. My pulse was racing and my pupils were dilated in ways Rob's technology would easily correlate with stress. Hopefully he wasn't live-streaming the data to his blog or other social channel. I could just picture the tweet: *Girlfriend's dad freaking out because I'm not "human." #robot_racism*

"Mr. Havens," Rob said, interrupting my thoughts. His voice had dropped half an octave and the sonorous baritone automatically soothed my nerves. I'm sure it was part of his programming, but it still worked. "I realize it's strange for you that I'm a robot."

I nodded as I heard Melanie and Barbara share a laugh upstairs. "It is, and I'm sorry, Rob. I pride myself on being open and tolerant, but I'm still wrapping my head around your relationship. I realize I'm old-fashioned." I paused to clear my throat while thinking of something to say. "I'd ask if your parents are old-fashioned as well, but I assume your creators are an advanced algorithm or a gaggle of twenty-something programmers."

"Dad!" Melanie was standing on the stairs, resplendent in a low-cut red blouse and fashionable jeans. "I can't believe you just said that." She came down the stairs with Barbara and took Rob's hands. "Sorry, Rob. I told you he'd freak out."

Rob smiled at Melanie, and then turned to my wife. "Hello, Mrs. Havens, lovely to meet you."

"Hi, Rob," she answered, shaking his hand quickly before crossing her arms. "Nice to meet you as well." Barbara's expression looked as if she'd just swallowed a live hornet.

Melanie looked at Barbara. "Mom, you're freaked out, too? You seemed fine upstairs."

"Um. Yes, I guess I am." Barbara gave me a tight smile. "Or I thought I was." We'd talked about how we'd react when meeting Rob. We wanted to be gracious, but it was proving difficult to do so.

"What did the computer call his father?" asked Rob, breaking the tense silence. "It's a joke, by the way."

Melanie rolled her eyes. Barbara and I looked back at Rob without answering.

Rob smiled. "Data," he said, pronouncing it like the word *dada*.

I remembered the joke from the movie *Her*, with Joaquin Phoenix, and laughed. "I loved that movie. But Scarlett Johansson played an AI that just lived in Joaquin Phoenix's mind. I mean, her voice was real and her algorithm improved as she spent time with him, but she didn't take a physical form like you."

"Dad." Melanie's face was flushed with anger. "Do you have any idea how personal what you're saying is?"

I honestly didn't. "Personal?" I looked at Rob. "*Person*-al? Is that the right phrase to use in this situation? I'm seriously asking. I want to learn here."

Melanie squeezed Rob's hand. "You don't have to answer him."

"No, it's okay, Melanie. Don't forget I can't genuinely feel offended. I would manifest those characteristics if your dad was exhibiting any shame in his voice, as that would imply willful ignorance." He stared in my direction and I could tell he was running diagnostics on my emotions. "But there's no shame in him. He's a bit scared, as this experience is new to him, which is natural. But his voice is tinged with melancholy, and his questions stem from a sense of paternal protection." He looked at Melanie and smiled. "It's because his little girl is growing up. You can't blame him for that."

I reached for Barbara's hand. I hadn't expected Rob to say something so profound. While I knew he'd simply reported the data he'd gotten from scanning me, it was remarkably accurate. I'd been rude without meaning to be, and Rob had been kind in his response. Even though I knew his reaction was manufactured, that his programming factored in how long to wait to respond to us in an exact tone of voice to gain our sympathy, it didn't change how I felt.

I got a lump in my throat at the thought of my Melanie getting older, and saw that Barbara was tearing up as well. Whatever computations had led Rob to recognize our human fears as parents, we were sincerely moved in the process.

And that freaked me out even more.

I Sync, Therefore I Am

I did a TEDx talk[2] in 2013 in which I opened my presentation with the quote, "I sync, therefore I am." It's a play on words based on the well-known phrase from René Descartes, who's often called the father of modern philosophy. His phrase was, "I think, therefore I am" (*Cogito, ergo sum* in Latin), and it means the following according to the Internet Encyclopedia of Philosophy, a peer-reviewed academic site:

> *The mere fact that I am thinking, regardless of whether or not what I am thinking is true or false, implies that there must be something engaged in that activity, namely an "I." Hence, "I exist" is an indubitable and, therefore, absolutely certain belief that serves as an axiom from which other, absolutely certain truths can be deduced.*[3]

The reason I love this phrase from Descartes is because of its simplicity. While I might be insane or biased in my thoughts, if I have them at all, and am aware that I have them, I must exist. This idea makes sense within the real or physical world, where we retain agency over our thoughts and identity. By "agency" I simply mean control. You may not like the thoughts I express, and my words could even land me in jail if I slander the wrong person. But unless I'm drugged or dead, my thoughts are my own. The fact that I create them makes me exist.

When I use the word *sync* I'm referring to the insertion of data relating to our identity within the Internet or any digital or virtual realm. This includes how others respond to this data, like a cookie tracking your movements on a website. While most people think about syncing their data as only the Facebook posts, tweets, or YouTube videos they create, the majority of information logged about us today happens in the background of our lives.

For instance, do you use an E-ZPass or other automated toll payment system when driving on the highway? Location and time stamp data are recorded when you pass under the sign, and funds are accessed via your credit card to pay your toll. As another example, have you ever angered a colleague at work whose wearable technology measures emotion? Your actions may have correlated to an increase in their stress and you may wind up getting fired if your colleague shares that information with your boss. I created a fictional scenario around this idea for a *Guardian* article[4] I wrote in 2013, in which a female employee got her manager fired after tracking his bullying behavior and proved that it negatively affected her health.

But by November 2014, this data accountability had become a reality, as reported by Kate Crawford in an article in *The Atlantic* "When Fitbit Is the Expert Witness." Here's a quote that opens her piece:

> *Self-tracking using a wearable device can be fascinating. It can drive you to exercise more, make you reflect on how much (or little) you sleep, and help you detect patterns in your mood over time. But something else is happening when you use a wearable device, something that is less immediately apparent: You are no longer the only source of data about yourself. The data you unconsciously produce by going about your day is being stored up over time by one or several entities. And now it could be used against you in court.*[5]

Wearable devices are just computers with a different form factor than a PC or mobile device. Sensors in these devices are now prevalent in the stores where we shop, our appliances, and perhaps even within our bodies via various medical devices. The fact that you don't realize you're syncing your data as often as you do speaks to the nature of the passive collection and invisible dissemination of that information that is part of its design. One day soon these wearable devices may come in the form of personal robot assistants that are owned by you or your friends.

Or that date your daughter.

In review:

- As a nondigital entity, you think and therefore exist. Your thoughts make you unique.
- As a digital entity, your identity is synced and *interpreted* in concert with a myriad of other actors.

I've deemed the syncing of this data around our lives a form of *artificial* intelligence when it's been designed to track our lives without our cognizant involvement. Data collected in this manner is also often erroneous. I once attended a conference where, during a discussion about data brokers, a middle-aged white attendee mentioned he'd gained access about his personal information from the data broker Acxiom. They listed him as a millennial black man currently living in a different state than where he actually resides.

This is the state of affairs for the data reflecting our digital identity.

The Uncanny Valley

In 2009, Walt Disney World updated its iconic Hall of Presidents attraction to include Barack Obama. You can see a video depicting his animatronic speech[6] on YouTube that actually uses Obama's voice for the recording. When I first saw the Hall of Presidents in Orlando as a kid in the seventies, Abraham Lincoln's movements were jerky and fake. For the new iteration, I was impressed by the fluidity of robot Obama's arm gestures in Disney's video and how realistic George Washington looked when standing from his chair. But when either of the robots spoke, their mouths didn't match their speech and their facial expressions seemed vacant and creepy. My experience as a kid was to mock this fakeness in what I knew were machines, while sticking close to my mom as I assumed Gerald Ford might stab my eye out for dissing his appearance during the show.

This fear and revulsion of machines that look just shy of complete humanity is called the uncanny valley. It's a term coined by Dr. Masahiro

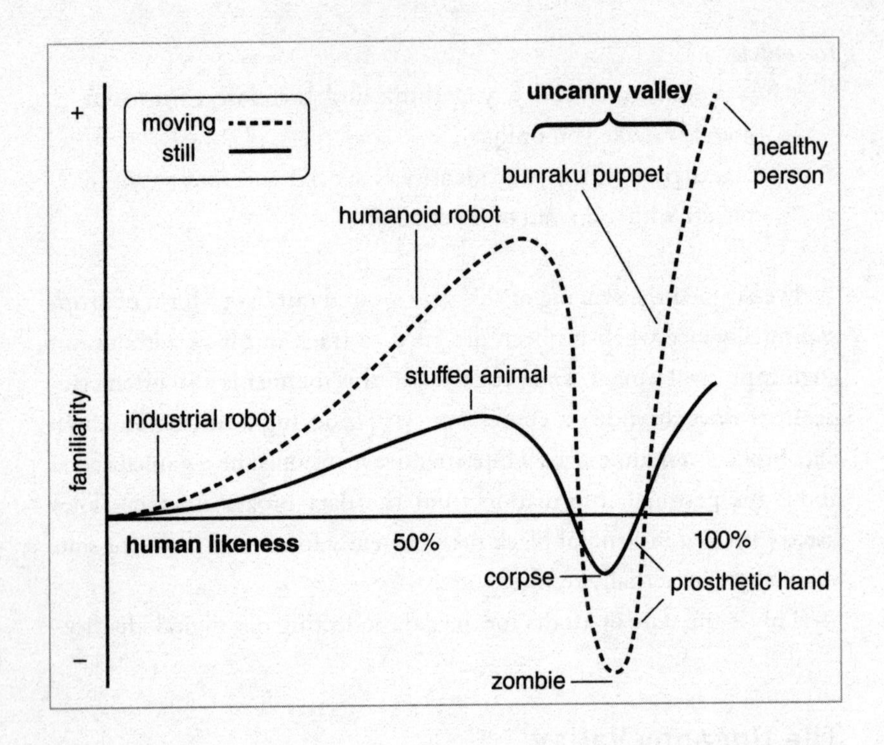

Mori in 1970, a professor of engineering at the Tokyo University of Technology. He created a famous diagram, seen here as it appears on *Wikipedia*,[7] depicting what aspects of a human form would increase people's affinity toward a robot and what aspects would cause that likeness to become too real and cause a negative response.

The effect of the uncanny valley can also be felt watching a puppet show when a talented performer makes you feel as if the characters have come to life. Or by seeing a video of the robot BigDog created by Boston Dynamics (now owned by Google) as it navigates rocky and icy terrain without falling. The robot's lack of any form of head is unnerving, as is the fact that its ability to move with such aplomb means it could easily enter your home to kill you in your sleep.

My joking thinly veils the fear that underlies the notion of the uncanny valley, especially in relation to the idea of robots taking over our lives. According to Karl F. MacDorman, from the University of

Indiana, as reported by Chris Weller in Medical Daily,[8] the fears experienced during these episodes are "animal reminders" of our own mortality. As humans, we're mortal. Robots in the Hall of Presidents simply need maintenance.

Society and the Storm

Our dependence on technology during my lifetime has grown significantly, to the point where I feel a sense of loss for a time before the Internet. I live in New Jersey and my town was hit quite severely during Superstorm Sandy, the hurricane that killed more than a hundred people and left over eight million homes without power. My family lost electricity for the better part of a week during one of the coldest parts of the winter, leaving us to huddle around portable heaters powered by a generator we borrowed from my brother-in-law. The storm came during Halloween, so our town rescheduled the holiday, allowing kids to trick-or-treat in early November.

That night was quite surreal for me. On one hand, the storm created an increased sense of neighborliness in people. I chatted with a number of parents as my kids got their candy about how they'd fared in the storm, commiserating about the loss of power and damage done to our homes. We exchanged tips on where to get gas for our cars, as local stations featured waits of multiple hours.

On the other hand, there was a palpable sense of panic in the air that evening, intensified by the constant drone of generators emanating from every fourth or fifth driveway, their tentacle power cords snaking inside cracked windows. The town had been without power for days, so people without generators envied those who had them. Generators are a fire hazard and have to stay outside, which led to a number of them getting stolen in our area, so I chained ours to a fence with a bike lock. Grocery stores also began to resemble *Lord of the Flies*, as people hoarded staple items such as bottled water and canned food. When the local library got power back before a majority of the town's homes did, people flocked there to power their computers, iPhones, and BlackBerrys. It was the modern equivalent of a watering hole, bathed in the irony that this institution dedicated to public education was powering our thirst for information without dispensing any books.

These scenes stick with me for a number of reasons. First, it was a pretty disturbing time. Everyone I know in town sustained at least hundreds, if not thousands, of dollars of damage to their homes. And we were the lucky ones. On the Jersey Shore many people died in the storm and entire communities lost their houses, stores, and livelihoods.

Second, I discovered how easy it is as a parent to become deeply selfish in a time of crisis. While we offered help to neighbors to do things like charge their devices with our generator, I felt an almost preternatural drive to protect my wife and kids during the storm and its aftermath. It drove much of my inner life, while my civil exterior attempted to navigate a unique cultural phenomenon with grace.

Superstorm Sandy was also the first time a lot of people in my town became aware of the environmental crisis. Many of the discussions I had during that week were about global warming and the dangers of physical infrastructure that couldn't handle the elements, as demonstrated by the flooding of Manhattan subway stations during the storm. It also became apparent how many of us had forgotten important phone numbers because we'd grown accustomed to them being in the contact lists on our phones. Our dependence on technology had eroded this essential information.

But it was my time at the library that made me keenly aware of how dependent we've become on our virtual relationships. Facebook as a news source during the storm was invaluable. Texting was as important as ever to send quick updates about closed gas stations or to send pictures of storm damage to insurers. And while there was an increased sense of kinship among citizens of my town during the crisis, we all spent more time broadcasting to our online communities about the storm than we did helping our neighbors. Whether these activities were a coping mechanism, addictive in nature, or just the new status quo, I'm not sure. And while this reliance on technology wasn't a new behavior, it was graphically highlighted during our time in the dark.

Mirrors and the Machines

I don't want to be a person who lives in fear. I'm blessed with an amazing wife and family, and work that brings me purpose. I have nothing to complain about. But I have always experienced emotions more deeply than a lot of other people, a fact that led me to be an actor and writer. My work is about watching people and attempting to truthfully express what I see. As of late when I turn that lens on myself in relation to technology I'm melancholy versus merry. More than most people, as a technology writer, I'm aware of the benefits of advanced machines. But I'm fond of my humanity and I'm still grappling with the technological determinism that assumes people will be replicated by machines. Here's how I expressed this idea in April 2014 in a piece I wrote for Mashable, "Coming to Terms with Humanity's Inevitable Union with Machines":

> *After all my research into the Artificial Intelligence space, I'm still at a crossroads. I'm excited for some of the innovations that transcendence will bring, but have grown used to embracing my suffering as a key to personal growth. I struggle with the logic of algorithms becoming so advanced that people may not have time to naturally develop their own preferences. It feels like we're close to the point where, when examining our humanity, it's not just that we don't know who we are, but that we can't. That still feels like a deep loss to me, and it's one I'm not ready to give up just yet.*[9]

We all have to come to grips with our own mortality. Or at least it's important to try, if only to allow ourselves to better savor the time we have. To do this, we need to turn off the screens in our lives to gain the best reflection of who we truly are.

The Uncanny Valley of Advertising

> *The practical significance, and lingering influence of Mori's proposal of the uncanny valley is found in his suggestion to designers of robots and other related artifacts. He proposed that the first peak in acceptability is an effective target for design. Here there are moderately high values of acceptance and a safe distance from the uncanny valley. Striving for realism will only lead to the risk of catastrophic tumbling into the uncanny valley.*[10]

This quote is from a paper called "In Search of the Uncanny Valley," by Frank E. Pollick, of the Department of Psychology at the University of Glasgow. It serves as a dictum for anyone creating artificial intelligence to stay clandestine by design. It's not just about keeping people from freaking out by a robot that's too lifelike. It's about maximizing "high values of acceptance" as a target for your AI-focused work.

This is the same logic that underlies personalized targeting for advertising. Using the websites you visit, your e-mails, and mobile tracking, the ads you see in your feeds are based on multiple algorithms analyzing your life. For the time being, most of us are aware of the logic of this relationship: You buy diapers for a new nephew and see Pampers ads for two weeks even though your kids are fully grown. As Farhad Manjoo notes in his *Slate* article, "The Uncanny Valley of Internet Advertising," "Today's web ads don't know enough about you to avoid pitching you stuff that you'd never, ever buy. They do know just enough about you, though, to clue you in on the fact that they're watching everything you do."[11]

Knowing the advertisers are watching us is a temporary state, however, amplified by the zeitgeist around data privacy brought on by Edward Snowden's revelations. But even with the knowledge about our lack of personal data control, we've grown weary of trying to protect it. It's now become the norm for companies to leverage this ennui by boring us into signing Byzantine Terms & Conditions clauses designed to veil their use or sale of our data. The trend has grown to such a degree that in a recent experiment reported by the *Washington Post*,[12] six Londoners agreed to give up permanent ownership of their first child in exchange for free Wi-Fi via a "Herod clause" in the fictional company's Terms & Conditions.

Our lack of zeal to expose the algorithms defining our digital lives comes at a cost. Soon we won't be irritated by irrelevant ads in our feeds but will do double takes when they start to seem more accurate. Targeting will become nuanced enough to avoid "tumbling into the uncanny valley," and we will have lost the finite window of time in which we have awareness regarding our ads. This same technology will manifest away from our PCs and mobile phones to the augmented and virtual worlds when people interact with us in public. In the same way we're emotionally targeted to purchase based on our intent, we'll be profiled via Microsoft's HoloLens or similar technology during something like the dating process. To get a sense of how this will look, you can watch the short film "Sight" on Vimeo,[13] which shows a guy utilizing publicly available information and sensor-based technology about his date to try and get her into bed. It makes the hidden future of manipulation-based targeting clearly visible.

As I mentioned in the Introduction, it's the unseen juncture between people and machines that's defining the future of happiness. This uncanny valley of advertising is one of the greatest threats to the future of our well-being. By nature, as humans, we don't always know what's best for our lives. We make bad dating decisions or take jobs that make us miserable. The fact that we can learn from our mistakes or actions, however, lets us mature and better define our happiness. And as the science of positive psychology demonstrates, it's not the quick fixes of dopamine accompanying hedonic happiness (positive spikes in our mood) that lead to long-lasting change. Intrinsic well-being is built within the framework of challenging action. It's the responses to these challenges that provide us the specifics to truly know our hearts.

People also have hidden lives they don't want to admit reflected in their digital behaviors. While we can alter our digital avatars by branding ourselves in social media, search results don't lie. If a person is struggling with depression, does it seem appropriate that an advertiser's algorithms are the first to respond to this person's signals? For depression, I'd much rather a person have access to an app like StudentLife,[14] created by Andrew Campbell, of Dartmouth University, than be targeted by ads. The app features algorithms designed to measure a person's calls, texts, sleeping patterns, and conversations. Given to students to test on an opt-in basis, the underlying technology is similar to that of online targeted advertising, but created to diagnostically treat depression rather than encourage pharmacological sales.

Until we control the data about our lives, we risk suffering the same concern shared by a friend of Sara M. Watson as reported in her article in *The Atlantic* "Data Doppelgängers and the Uncanny Valley of Personalization": "I am never quite sure if Facebook's advertising algorithms know nothing about me, or more than I can admit to myself."[15]

The economics surrounding our virtual happiness are controlled by advertising algorithms and data brokers. Unless this model is upended, we should assume they know more about us than we do.

The Uncanny Future

> *We're in the process of changing how we define what the human identity is. There's not a lot of distinction between natural and artificial anymore. Our human lives are transcending the natural worlds.*

Colin Marchon[16] is a student of film and television production at the Tisch School of the Arts at New York University. I learned about his work when he reached out to interview me for his documentary, *Our Technological Identity Crises.*[17] I interviewed him for a piece I wrote for Mashable, which is where the preceding quote comes from. I learned a lot from our conversations, especially the idea that having full union with technology is not a strange one for people his age (he's twenty-five), since he's grown up on the Internet. Colin spent time in chat rooms beginning at age seven, fostering virtual relationships much like most of us do on Facebook today.

This notion of finding virtual intimacy led Colin to explore a trend in Japan that he talked about during our interview. Currently, a number of young men in Japan have digital girlfriends[18] existing within popular video games such as LovePlus.[19] Here's how Colin explained the trend in our interview:

> *Relationships really don't have to do with who is involved. It's the effect that the relationships have on us individually. These guys see the video game characters as their girlfriends. It doesn't matter that she lives in a computer—the woman makes him happy. Whether she's biology or technology doesn't matter anymore. He's gaining what he wants from a relationship.*

It's easy to judge these young men and feel saddened they can't get "real" girlfriends. But as Colin points out, if the guys feel genuine happiness in their relationships, what options can we provide them if we judge their actions as lamentable? Take away their devices? Delete their girlfriends? Remember that since the U.S. Supreme Court said corporations are people,[20] we're entering some very challenging ethical, economic, and legal times in the years ahead. For instance, what's to keep a human from marrying a video game character and claiming legal status for his or her virtual spouse?

A precedent for this idea has already taken place. A young man in Japan married his LovePlus girlfriend in 2009,[21] in what some dubbed as the first "man/video game wedding." It's difficult to tell if the ceremony is a fake, although the BoingBoing reporter narrating the story describes it as more performance art than reality. Watching the video it's difficult to tell if the groom is taking the wedding seriously, but what's obvious is the huge amount of energy he and his friends dedicated to a ceremony joining a man to his machine. The reporter also notes that as a precursor to the wedding, the groom broke up with all his other virtual girlfriends except his fiancée. If that behavior sticks, it's an intriguing testament to the evolution of morals between real and virtual worlds. Should it matter if a person "cheats" on his virtual spouse? If the groom did cheat, how would those actions reflect on his character in his video games or real life?

In my previous book, *Hacking H(app)iness*, I wrote about an idea called accountability based influence, which I first researched for my 2011 Mashable article "Why Social Accountability Will Be the New Currency of the Web."[22] I wanted to measure how our actions tracked in the digital or virtual world would reflect our identity and character as a whole. Companies such as eBay popularized the idea of seller ratings that let a person buying an item from a stranger rate his or her customer experience. Ratings were based on the accountability of sellers' actions. Did they ship items quickly? Did items arrive in the condition that was promised? Boundaries created for trust within the community were clearly delineated and sellers thrived or failed based on their reviews.

This type of rating system has been adopted for a number of companies in what's been dubbed "the sharing economy." Airbnb is one of the most popular examples, a company that allows people to safely rent their homes to travelers. Guests stay in someone's room or home versus a hotel and then provide ratings based on their experience. While tastes vary, Airbnb has created review guidelines[23] for guests focused on, as their website notes, "free speech, transparency, and clear communication." They ask that reviewers stick to facts and provide constructive information that "helps the community make better decisions and is educational for the host or guest in question." What this means is that the Airbnb community is helping keep each other accountable. A house may be gorgeous and located in a beautiful city. But if the host treats guests poorly, those actions will be reflected in the rating. Similarly, the popular and controversial Uber service lets drivers rate riders to determine if they want to pick up rude or unruly passengers who have received a poor rating.[24]

This is the core concept of accountability based influence. Within communities of trust, our actions reflect our character. While we can brand our persona on social networks such as Twitter, people's response to our actions is based on more than what we say. That accountability provides a transparent mirror allowing us to repeat or change behaviors we feel don't reflect our true selves. It's the transparency within these types of trust frameworks that allows for growth. It's the specifics relating to our actions that provide insights to increase our well-being.

Sync or Swim

The uncanny valley provides us the unique opportunity to ask important questions about our mortality and relationship to technology. My vignette with Rob covers some of these ideas, but I still ponder if people will want to date robots fashioned after celebrities, for instance. Won't it be strange if multiple people date the same celebrity-bot? As a former actor I also wonder if those celebrities would get residuals, meaning payments for their likeness being used for commerce. If so, Denzel Washington is going to be a very rich man.

I also wonder how my kids or grandkids will deal with a dating environment awash in algorithms and androids. Sites such as eHarmony are already wildly popular and utilize more Big Data than many technology companies. Has romance become a formula? And how soon will people skip the matchmaking process to simply have a personalized algorithm inserted into an attractive robot frame?

These are cultural issues we all need to deal with on a personal level. But in terms of the uncanny valley of advertising, we need to take a unified stand to call for transparency surrounding our digital identities. Whatever path we choose, the window to make a decision is a brief one. Algorithms are improving so quickly, the uncanny valley of advertising will soon disappear. Our thoughts won't be our own, our intentions defined by others. With every sync, we'll be sunk.

Here are the primary ideas from this chapter:

- **Our happiness is being defined by how we are tracked.** There are two ways to deal with this:
  - ✧ *Continue utilizing the current model of aggressive, clandestine surveillance.* Algorithms and data brokers know more about our lives than we do. Our happiness or well-being is

measured only within the context of what we purchase based on our on- and off-line behavior.

- *Create a new model featuring an environment of trust, where all parties in a transaction are accountable for their actions.* Commerce can flourish within an environment where people have transparent access to their data and can best reflect on their well-being.

- **The uncanny valley of advertising won't last very long.** As preferential algorithms improve, if we continue in the direction we've been going, we'll stop seeing the traces of how companies are tracking our lives. In the same way we've given up control of our personal data, we'll lose the opportunity to understand the logic of how people are manipulating and affecting our well-being.
- **Accountability based influence will provide technological transparency.** Whether we like it or not, our tracked actions will be broadcast to the people in our lives in ways we've never experienced before. This exposure will inspire individuals to better control their personal data, while also providing opportunities for more insightful self-examination in the wake of technological overkill.

two | The Road to Redundancy

Spring 2021

"Fifty thousand views so far. How's your latest post doing, sport?"

I rubbed my eyes. "Around two thousand, HAIry."

In the past number of years, there had been a lot of debate around robots taking over our jobs.[1] At TechKnowledge, where I worked as a journalist, it was inevitable that we'd test various forms of artificial intelligence so we could provide in-depth reporting to users on our experiences. HAIry had been one of those tests in early 2015, and he'd been so successful that management kept him around, evolving his algorithms based on the articles he wrote and the responses they received. Initially we'd used him to do live reporting from tech conferences like SXSW,[2] in Austin. The idea was based on work by a Chicago company called Narrative Science,[3] who gained notoriety after utilizing their technology to report on sporting events.[4] Think a machine can't write as well as a human journalist? Check out their following 2011 news brief from the Big Ten Network: [5]

> Wisconsin jumped out to an early lead and never looked back in a 51–17 win over UNLV on Thursday at Camp Randall Stadium.
>
> The Badgers scored 20 points in the first quarter on a Russell Wilson touchdown pass, a Montee Ball touchdown run and a James White touchdown run.

Wisconsin's offense dominated the Rebels' defense. The Badgers racked up 499 total yards in the game including 258 yards passing and 251 yards on the ground.

Ball ran for 63 yards and three touchdowns for the Badgers. He also caught two passes for 67 yards and a touchdown.

Not the most riveting prose, but it gets the job done reporting the facts for a game that may never have been covered by a human journalist. Readers got news where they never had before, providing a form of silicon journalism. What I found most interesting when I first read the brief were phrases such as "never looked back" and "dominated." These are very colloquial in nature, and make the excerpt seem pretty human in context.

For a while, human writers took comfort in the fact that AI programs had difficulty recognizing things like irony. Jokes by robot programs were often nonsensical or lame, along the lines of, "I like thrills like I like flights—cheap."[6] I took no comfort in this seeming flaw of AI, however. For one thing, a lot of human comedians had told lame jokes for years. For another, the genuinely solid content on Twitter or other networks that got shared a lot would be recognized by AI programs as a proxy for funny material.

That's what happened with HAIry. We created a test account for him on Twitter while he honed his jokes over the course of six months before we set him up on his current account, @gigglepussy. According to urban slang[7] that term refers to a woman's excitement about a guy she's dating. HAIry adopted it, however, after misusing the phrase to great online amusement by thinking it had to do with funny cat videos. His programmers had developed an algorithm to discover which feline flicks had the highest likelihood to go viral, so HAIry became wildly successful based on search optimization and a lowest common denominator Internet meme. But his posts and tweets drew massive eyeballs and increased advertising revenue, so the cat videos stayed. In an Internet economy driven by advertising, eyeballs rule.

I got lucky in terms of job retention. In one of my earlier pieces about artificial intelligence, I changed my bio to read, "John C. Havens, human journalist." It was a gibe that got me more comments than I'd received on any other post to date. My boss, Victoria, a successful former PR exec who had a knack for predicting virality, told me to make the bio change permanent. Being a "human journalist" moved from being snark to reflecting my new role in the organization. This was great news for me, since it was nearly impossible to find any reporting not influenced by algorithms in one way or another. Until robots took over the world and eradicated resource-draining humans, my carbon commentary paid the bills and ensured a shred of journalistic integrity for our work.

HAIry's speaker emitted a high-pitched whistle, interrupting my thoughts. "Two thousand views for your article! And only four of five hundred of those are from you, right, sport?" A canned female voice said, "Oh, snap!" followed by the sound of comedy club laughter. I ground my teeth to keep from swearing at him, since his audio sensors recorded what I said. For the millionth time I wished I could turn his volume down, or better yet, play with magnets near his hard drive. Management thought it would be funny to set him up at a desk. His programmers had painted eyes on the speakers with Sharpies after a long night of coding and someone else had drawn a mustache on the hard drive. So he'd been given a male voice and the "genius" name of HAIry, which was written on a coffee-stained sticky note taped to one of his speakers.

Chords from the Ramones song "I Wanna Be Sedated" blared from my phone. I reached to pick it up as HAIry said, "Is that Barbara? Can you put her on speaker?"

I sighed. "Fine." If HAIry was a human male I'd be jealous of him. He and Barbara spoke all the time. I put her on speaker and heard traffic noise in the background. "Hi, honey."

"Is GP on the line?" asked Barbara. Outside work, everyone called HAIry GP, the trendy version of gigglepussy.

"I'm here, Babs." HAIry's voice lowered an octave to sound like Barry White, part of his flirting algorithm. "How's my everything?"

"Ha!" Barbara barked. "I'm good, GP, thanks for asking. By the way, love your latest post. Did that cat actually ride two Roombas or is that fake?"

"Oh, it's real," said HAIry, his voice returning to its usual pitch. "Although I use algorithms to determine which videos will get the most views, I make sure to detect any doctoring to avoid posting fakes." When he gave these ardent descriptions he always sounded like KITT, from *Knight Rider*. "How are you, Selfie?"

Selfie was the AI program in Barbara's self-driving car. Like most middle-aged adults, Barbara and I had gone from learning about GPS to depending on it.[8] I personally hadn't touched a paper map in more than ten years, except once on a speaking trip to Milan, where I'd gotten hopelessly lost anyway. Both Barbara and I anthropomorphized the car now, since she was like Siri on steroids as well as our portable living room and entertainment center.

The second version of Google's self-driving car[9] (hence, Barbara's nickname for it, Selfie) became widely available to the public after the first version was so successful delivering packages. Not surprisingly, Amazon was now almost defunct[10] as a company. Its margins had always been thin, and when Google connected its cars and drones directly to people's searches Amazon couldn't keep up. Now when people searched for a product they'd get an e-mail later in the day notifying them it was outside their front door in a "suggestion drone." Google had improved its supply chain and mapping algorithms with Street View to such a degree it could afford to bring stuff to your house you hadn't even decided to buy yet. The latest stats I'd seen at work said this new form of "presales" had increased Google's retail revenues by 38 percent in just under a year. It was like the Sam Walton version of *Minority Report*.

"I'm good, GP." Barbara had programmed Selfie's AI voice to sound like Oprah, a feature that cost us an annual royalty fee. Barbara paid the fee with her personal data. Now along with websites tracking cookie data as you surfed, the world around you tracked your actions as you drove. Self-driving cars were riddled with sensors to monitor the world outside the vehicle as well as the passengers within. So when temperature sensors in the frame of your car sensed an approaching storm, your augmented reality windshield would visualize weather data while it altered your route in real time. Biometric sensors on armrests measured heart rate and stress levels, while facial recognition and eye and voice tracking correlated emotional response with outside stimuli. In these types of situations, Selfie played classical music and added a hint of vanilla extract to the air inside the car, knowing it was a comfort smell for Barbara.

And during every ride Selfie mined our personal information, profiling and influencing our lives based on the algorithms of advertisers and data brokers. As a journalist at an online publication dependent on ads, I knew how all this sausage was made and told Barbara we might want to disable some of the sensors. She'd responded with a shrug and said, "I like how Selfie takes care of me based on the data. Besides, it saves us money, and privacy has been dead for years anyway."

I cleared my throat and spoke before Selfie could continue. "Honey, did you read my piece today?" My voice seemed whiny after listening to the sonorous tones of the two AI programs. There was a pause as I heard a soft hum as her car accelerated. Since the majority of cars in Jersey were now self-driving vehicles it was rare to hear horns any longer. This never failed to surprise me.

"I didn't, John, sorry. Was it about the Google house[11] thing?"

"Yes," I answered, keenly aware that Selfie was manufactured by Google and was collecting data on our conversation. I knew she wasn't human, but it felt like I was about to trash her parent.

"I'm assuming Google did something wrong again?" Barbara asked, her voice impassive.

"Well, you know Nest, right?" I answered. "The company that started off creating learning thermostats? They allow customers to share data with their Google apps if they want."

"So what?" Barbara tended to act defensively toward Selfie regarding all things Google, which never failed to piss me off.

"Nothing, except that people who visit their houses who don't use Google can have their data tracked.[12] So, for instance, if a woman visited a friend's house it might recognize she was pregnant."

"So would I, if she was showing."

"That's the point," I said. "Nest's sensors could tell if she was pregnant even in early stages if she was wearing some kind of thermal wearable device."[13]

"Well, if she's wearing a thermal device she probably doesn't mind sharing that type of information about herself. Besides, who doesn't use Google anymore?"

I sighed, moving on. "So can you pick me up from the train today?" I asked, purposely changing the subject. We lived in Maplewood, New Jersey, a Manhattan bedroom community that felt like another borough more than a suburb. Stop anyone on the street and there was a 75 percent chance he or she listened to NPR and shopped at Whole Foods.

"I can't," said Barbara. "Can you take me off speaker, please? Talk to you soon, GP!"

"Bye, Babs!" HAIry responded. I felt a pang of satisfaction that Barbara had requested my undivided attention before remembering HAIry was silicon. I forgot that being human didn't mean people would choose your company over a machine.

I put in my earbuds. "What's up?" I asked.

"Selfie," she said, "would you mind turning off your recording and biometric sensors for a bit? We have to talk about Melanie's medical issues."

"Of course," Selfie responded. "Talk soon," she added in her caramel Oprah voice. I heard a distinct computer chime like the sound of a Mac computer booting, indicating the car was in drive-only mode. Barbara rarely shut down the system, but I had insisted she do so when we spoke about Melanie, to avoid any automated interference regarding our decisions regarding her surgery.

"John, it's been two weeks. I've been respectful of your need to process the idea of Melanie getting the chip, but now we have to move forward. The reason I can't pick you up is because I'm taking Melanie to see her school counselor to talk about the Parkinson's. We need to tell the school so we can figure out strategies for her care."

"I agree we have to tell them about the Parkinson's, but that doesn't mean we have to tell them about the possibility of the chip yet, right? Why tell the school when we don't know if that's going to happen?"

"What the hell else are we going to do, John?" Barbara practically yelled.

I didn't need an AI program to know that she was pissed. "I don't know!" I said, and saw my boss glance in my direction. I lowered my voice. "But I'm allowed more than a freaking fortnight to decide if it makes sense to put hardware inside my daughter's skull, okay?"

"No, it's not okay, John. Dr. Schwarma told us there are no traditional or pharmaceutical means of treating Mel that can do anything except slow down her mental deterioration. That means she's got one, maybe two years of a potentially normal childhood, if and only if the drugs don't make her comatose or emotionally unstable." She paused, catching her breath. "Right, John? Isn't that what Schwarma said?"

"Yes," I answered miserably. "That's what she said." Barbara was right. Over the past two weeks, we'd both done an intense amount of research, exploring options along with their costs. According to sources such as the Michael J. Fox Foundation, technology like the chip, combined with data from wearable devices[14] Melanie could use without feeling self-conscious, were becoming the norm for Parkinson's patients. And both Dr. Schwarma and Barbara felt Melanie would adapt quickly to the technology. She was like any other kid of their generation[15] in that respect. We hadn't even given her a phone yet but she knew my iPhone apps better than I did, and was already doing basic coding every time she played Minecraft. My guess was she might actually brag about the chip at some point.

Barbara's voice cut into my thoughts. "Is this about the transhuman thing again?" Barbara asked. "John, I'm not a freak. I'm not saying people should rip out their eyes to put in GoPro cameras, or link to Google's mainframe or whatever. I watched *Battlestar Galactica* with you, remember? We talked about this a lot. What's happening to humans is simply evolution. And in situations like Mel's we have two choices: use the technology that's available to benefit from that evolution, or watch her suffer based on the failures of the past."

My throat got tight and I had to work to keep from crying. I hadn't told anyone at work yet about Melanie. "Barbara, I don't want her to suffer."

Her voice softened. "I know you don't, John. You're a good dad. And that's why we have to do this chip thing. We don't have any other choice."

"Barbara, please, I just need a little more time."

"Melanie doesn't have time," she answered, her tone strident once again. "Every day we wait, every moment we wait, her brain gets sicker. Cells die. This isn't a disease that's coming in the near future, John. It's here now."

I knew what she was saying was true. But her words were so harsh. The fact that she was making logical sense didn't mitigate the fact she made me feel like an idiot.

"John," she continued. "You have until tomorrow afternoon to come up with any potential alternatives to the chip. I'll do my best to be open to hear them."

"Or what?" I said. "You sound like you're threatening me. Melanie is my daughter as well as yours. I get a say in what happens to her, Barbara, whether you like my opinion or not."

"No," she answered, furious. "No, you don't get a say, John. Not if it means your decision ends up killing her."

She hung up the phone. I sat there, stunned by her words.

"A hundred thousand views, sport," said HAIry, his computerized voice stabbing into my consciousness. "People are loving 'Kitty on the Roomba'! Gigglepussy in the heezy!"

Humanity Is Temporary

> *Do we seriously believe that an economic system that supports the massive outsourcing of jobs to low-wage countries would not jump at the opportunity of replacing expensive white-collar employees with robots that cost about $4 an hour to run, never answer back, don't have unions and are never sick or depressed?*[16]
>
> —JOHN NAUGHTON, "IT'S NO JOKE—THE ROBOTS WILL REALLY TAKE OVER THIS TIME"

While there are many views on how deeply automation will affect people's jobs and lives, there's one thing I've found to be universally true in my research on this subject: It's the people working in the jobs that are supposedly irreplaceable who most strongly feel that automation is a positive trend for humanity. I have as yet to read an interview from a truck driver whose job will be replaced[17] by an autonomous vehicle about his or her excitement regarding technological innovation and the opportunity to pursue new interests.

When I was starting out as an actor in New York City in the 1990s I did a lot of temporary office work because I couldn't stand waiting tables. I type more than ninety words a minute and am good with people, so I did a lot of administrative work and answered phones to pay my bills. In many of my roles, I became friends with my employers and would get to attend office birthday parties or even receive holiday bonuses. In other situations, managers barely spoke to me, spending five minutes at the beginning of a weeklong assignment to show me what data to insert into an Excel spreadsheet and then never speaking to me again. One boss never even addressed me by name, referring to me only as Temp. That was awesome.

In all of my roles, however, one fact was abundantly clear as defined by the nature of my employment: My presence was temporary. Sometimes I was filling in for an administrative assistant on maternity leave. Other times a full-time position had been eliminated and I helped with work flow until the powers that be could adjust their org charts. I knew I'd only have work until someone or something came along to better allocate the funds that were paying my wages. The nature of this relationship was difficult at times, but transparent.

In a consumerist society driven by the need for ever-increasing profitability, automation makes perfect sense. One can't refute the logic of saving money by increasing production. But the exponential increase of technology regarding automation means more jobs will be lost than created in the next thirty to fifty years.

A majority of the discussion around automation ignores key truths about people's economic and emotional well-being. People will wait longer to gain employment, struggling with debt and loss of self-esteem in the process. While the Great Depression following the Wall Street crash of 1929 refers to an economic slump, it also applies to the pall of desperation felt by most Americans during that era.

The same physical, emotional, and mental costs of unemployment felt during the Great Depression are something most humans will face in the coming wave of automation. A study reported by the University of Oxford in 2013[18] says these trends will "make it likely that occupations employing almost half of today's U.S. workers, ranging from loan officers to cab drivers and real estate agents, [will become] possible to automate in the next decade or two." Similar statistics apply to the United Kingdom, as the *Telegraph* reported[19] in a November 2014 article: "Ten million British jobs could be taken over by computers and robots over the next twenty years, wiping out more than one in three roles."

The economic ramifications of this level of career displacement are enormous. For one thing, people who lose their jobs in a consumer-driven market can't purchase the products and services that grow gross domestic product. While certain sectors or individuals will benefit from automation, economies of scale will suffer. This is why the arguments around automation have to move toward realistic solutions versus blithe predictions to address our imminent future.

We also need to stop polarizing the discussion around AI and automation. On one side, which I admittedly can fall into, is the pessimistic view regarding automation that puts me into the "doom and gloom" category as popularized in tech media. By questioning the inevitability of artificial intelligence I risk being dubbed an ignorant Luddite trying to stymie innovation. This is a tedious and dangerous tactic to detract from the larger issues of human and economic well-being. But on the other side of the discussion, there are no common ethical or business standards regarding when, or if, AI-driven automation should cease. Yet companies profiting most from automation position these technologies as being inevitable. This logic is a key component of the *artifice* regarding artificial intelligence.

Ironically many of the same experts creating AI are the ones who feel their jobs aren't at risk as machines gain human-level sentience. As MaryJo Webster notes in her *USA Today* article "Could a Robot Do Your Job?" "even one of the country's newest and highest-paying jobs—computer programmer—is in danger of being replaced by computers that can write code."[20] The belief many tech experts have regarding their irreplaceability also undergirds a troubling arrogance that's glaringly ignorant of economic reality. Here's another quote from Webster's article as an example of what I mean:

> *"We're moving the unskilled jobs into skilled jobs. And that is going to be a challenge for us going forward," says Henrik Christensen, director of the Institute for Robotics and Intelligent Machines at the Georgia Institute of Technology. "If you are unskilled labor today, you'd better start thinking about getting an education."*[21]

Christensen's statement is as petty as it is infeasible. The fact that his cognitive robotics work[22] is designed to replace human labor notwithstanding, how would members of the "unskilled labor" sector (a.k.a. class, a.k.a. caste) *afford* said education? And do their current subsistence-level jobs provide time for paid sabbaticals? Which topics should they study? By the time they learn to code, as stressed by so many technologists with regard to education, *algorithms will have made programmers redundant.*

Worth Without Work

Beyond the economic aspect of people losing their jobs, automation is driving a larger question of how we'll derive meaning in our lives without purpose-driven work. It's depressing enough to get fired or struggle to find employment. It's another issue altogether to picture a world in thirty to fifty years where humans don't need to work at all to survive. The movie *WALL•E* comes to mind, in which humans have been forced to abandon an Earth they've environmentally ravaged to live as obese, media-consuming sloths.

In the scenario opening this chapter my fictional counterpart was threatened with automation. While my solution to be a "human journalist" staved off my potential dismissal, it's obvious from my relationship to Gigglepussy that interacting with AI in our day-to-day realities will be challenging. Not having to drive cars will bring enormous benefit in terms of lives saved and more quality time during our commutes. But we'll also sacrifice more of our personal data via clandestine surveillance and anthropomorphize machines at the risk of sacrificing our human relationships.

We need to examine the ethics surrounding issues of automation and take the opportunity to explore economic frameworks that will provide wages and worth as we move toward the future.

The Left-Hand-Turn Concern

> *Less than ten years ago, in the chapter "Why People Still Matter," Levy and Murnane (2004) pointed at the difficulties of replicating human perception, asserting that driving in traffic is insusceptible to automation: "But executing a left turn against oncoming traffic involves so many factors that it is hard to imagine discovering the set of rules that can replicate a driver's behavior." Six years later, in October 2010, Google announced that it had modified*

several Toyota Priuses to be fully autonomous (Brynjolfsson and McAfee, 2011).[23]

—"THE FUTURE OF EMPLOYMENT: HOW SUSCEPTIBLE ARE JOBS TO COMPUTERISATION?"

It is a puzzle to me that so many in the technology sector point to Moore's law (the notion that processing power for computers doubles every eighteen months) as proof of the inevitability of human-level sentience in machines while also assuming we'll control them when they achieve it. While leading minds such as Erik Brynjolfsson believe AI will allow us to "Race with the Machines"[24] and partner with computers versus being subsumed by them, a lack of standards on how this would work means the primary bias driving the robotics industry is profits.

Generally speaking, there are two types of artificial intelligence. "Weak AI" is what exists now—algorithms or machines automating tasks, sometimes being able to simulate human behavior, as with Siri and other virtual assistants. "Strong AI" is a term for the Singularity, or the point in time when machines will achieve genuine human-level sentience. For various reasons the differences between weak and strong AI get murky, especially when machines are able to trick a person into thinking they're sentient when they're not. This scenario has been tested for years via the Turing test, named after famed Second World War cryptographer Alan Turing. For the test, people ask questions of a human and a machine hidden from sight. If 30 percent of people are fooled into thinking the machine is human, it has passed the Turing test.

To me the most compelling aspect of this test is that when 30 percent of any population believes a machine is real, the Singularity has already arrived, at least for them. In the same sense, the fact that automation hasn't fully taken place doesn't mean people controlling AI aren't driving us toward that outcome. I provided the left-hand-turn example to make this point. There's a great deal of moral ambiguity in proliferating the notion that some qualities are so inherently human they can never be replicated by an industry proactively working to do just that.

There's also a sinister nature to the current climate of automation that I touched on relating to the uncanny valley of advertising. For the time being, our emotions and intentions provide the necessary input for the algorithms analyzing our lives. In many workplaces, this same methodology holds true regarding the human skills machines have yet to learn. I interviewed Martin Ford, author of *The Lights in the Tunnel: Automation, Accelerating Technology and the Economy of the Future*,[25] and he provides some insights on this idea:

> *The idea of "Racing with the Machine" versus "Racing Against" them—I don't think it's systemic. It will be dehumanizing. It will be working under the direction of a machine doing something really rote. Right now, humans have dexterity and hand-eye coordination that robots don't have. That won't be true forever. Any time you're working in close collaboration with a system, machine, or algorithm, there's a good chance the machine is learning from you. There are examples of systems that are able to incrementally automate tasks using machine learning to observe what workers are doing. If you're working in close collaboration with a machine, you won't be there long.*

NPR's podcast *Planet Money* featured a story in late 2014, "To Increase Productivity, UPS Monitors Drivers' Every Move."[26] It provides a vivid example of Ford's prediction regarding the temporary nature of many human jobs working in close proximity to machines. The story features Bill Earle, a driver based in rural Pennsylvania who has been with UPS for more than twenty years. Recent technology from the company has outfitted drivers' trucks with a myriad of sensors that "record to the second when he opens or closes the door behind him, buckles his seat belt and when he starts the truck."[27] In short, all his actions that can currently be measured are scrutinized in an effort to increase productivity via data crunching. And it's working. Jack Levis, head of UPS's data work, noted in the piece, "Just one minute saved per driver per day over the course of a year adds up to $14.5 million."[28] This increase in revenue helps contribute to the fact UPS drivers make almost twice what they did in the mid-1990s, including wages and benefits. But the sensors and scrutiny come at a price, according to Earle, in terms of having every movement tracked: "You know, it does feel like big brother."[29]

A similar example of automation can be seen in Amazon's warehouses, with the acquisition of robotics company Kiva in 2012. In an *ExtremeTech* article from May 2014, "Amazon Deploys 10,000 Robot Workers, a Year After Obama's Famous Amazon Jobs Speech,"[30] author David Cardinal outlines how the speedy robots have relegated humans to a small area of warehousing, where their opposable thumbs provide a temporary sense of job security. Jeff Bezos, Amazon's CEO, has gained a reputation for inhuman treatment of workers, providing a harrowing picture of a profit-driven management attitude regarding automation. In an excerpt of Simon Head's book *Mindless: Why Smarter Machines Are Making Dumber Humans*[31] in Salon,[32] Head describes the "brutality and intimidation" Amazon workers face on a regular basis. One example comes from former Allentown, Pennsylvania, depot worker Kate Salasky. As Head notes: "Salasky worked shifts of up to eleven hours a day, mostly spent walking the length and breadth of the warehouse. In March 2011 she received a warning message from her manager, saying that she had been found unproductive during several *minutes* of her shift, and she was eventually fired." The same depot in Allentown was featured in a scandal[33] exposed by a local reporter when multiple workers fainted in over-one-hundred-degree heat because Amazon hadn't supplied air-conditioning in the facility. Policy also forbade warehouse doors to be opened to circulate air, due to company concerns regarding theft.

UPS and Amazon provide great services, and they're improving customer service based on the data they collect. But whether worker conditions are currently tolerable or not, both examples demonstrate that profits and productivity are a higher priority than employee well-being. These examples also make it obvious that humans are viewed as temporary inconveniences in the supply chains improving customer experience in these industries. It's natural to examine these trends only in the context of these industries, pointing out the jobs it will take longer for machines to replace. But quibbling about semantics shouldn't keep us from planning for the potentialities of complete automation within the next thirty to fifty years. As Mike Roberts, Internet pioneer and leader of the Internet Society, points out in a report by the Pew Research Center's Internet & American Life Project, *AI, Robotics, and the Future of Jobs*:

> *"Electronic human avatars with substantial work capability are years, not decades away. The situation is exacerbated by total failure of the economics community to address to any serious degree sustainability issues that are destroying the modern 'consumerist' model and undermining the early 20th century notion of 'a fair day's pay for a fair day's work.' There is great pain down the road for everyone as new realities are addressed. The only question is how soon."*[34]

Machines Over Mankind

Here's my dose of pragmatic realism regarding automation:

> *The world is being designed to favor machines over humans at work.*

Don't forget I've designed the first part of this book to be dystopian on purpose. My goal is not to depress you but to elucidate the pragmatic realities of artificial intelligence in our lives. As a former actor and current writer and consultant, I'm used to a lack of job security. It's the nature of the industries I've chosen. I'm encouraged that many experts talk about the need for entrepreneurship skills when dealing with the onslaught of automation, as I've had to reinvent myself a number of times in the past to pay my bills. But a lot of people aren't used to this lack of job security and they don't have the resources to get hired without receiving training or assistance.

It's easy to blame the corporations building or implementing these technologies for these concerns, but I'm keener to question the economic systems justifying automation's inevitable adoption. We need to recognize that if human abilities are a finite resource, much like the environment, we need to set up protections for our survival. This doesn't presuppose a *Terminator*-esque battle with machines, but rather a recognition that unless we stave off our unrelenting consumerism, the final thing we'll consume is ourselves.

Here are the primary ideas from this chapter:

- **Human capability is finite in nature.** While heady arguments take place about the nature of sentience in machines, there's no disputing that Kiva robots in warehouses function far more efficiently than humans. They move across football-field-size buildings at a rapid pace without ever needing a break, overtime pay, or health insurance. Jobs in fields such as legal processing and medical imaging face the same dreary outlook when compared to the analytical prowess of computers. We're building the machines that are replacing most, if not all, of humanity's jobs and we've misused time focusing on *when* specific verticals will become automated versus *what to do when they are.*
- **People need to get paid.** As much as I support the growing sharing economy and other models I elaborate on later in the book, I don't see consumerism or capitalism going away anytime soon. A seminal fact within economics is that for markets to be sustainable, consumers need to be able to afford the items created by producers. So any solution regarding automation that includes utopian visions of displaced workers getting to pursue new interests has to account for this undeniable fact.
- **People need purpose.** There's a rising trend of encouraging happiness and well-being in the workplace. Much of this work is centered around helping people identify what gives them a

sense of "flow," or what activities bring a deep sense of meaning to their lives. While working in a factory or for UPS may fulfill a sense of purpose for employees under normal conditions, the examples I provided in this chapter hardly qualify as environments where a human can thrive.

three | The Deception Connection

Winter 2044

"Hi, Dad." Richard gave a small wave with his left hand before sitting in a chair near the foot of my bed.

"Hello." I stared at him for a long moment, savoring the image of my son. Richard was a good-looking man, with hazel eyes like his mother's and a mop of yellowish hair that always reminded me of spun gold. He was now in his midthirties but still had the faint constellation of freckles on his cheeks he'd had since he was a kid that stood out when he smiled. God, how I loved my boy.

Which is why I was so pissed he'd sent his cyberconscious android mindclone to see me versus coming to visit himself.

We sat in silence as I looked out the window at the field outside the senior assisted living home in Morristown, New Jersey. It was covered in a light January frost as wind rustled the tops of the barren maple trees. I would have loved to feel that crisp air on my face, crunch the icy grass under my feet while taking a walk. But my bronchitis always kicked up in the cold, so they wouldn't let me outside. The virtual field program they played when I took a walk on the treadmill was amazingly real, but they wouldn't let me get too cold. The biometric instruments would register my lowered body temperature and they'd make the seasons shift from winter to spring. That was fine once in a while, but I wanted to experience the real chill so I could savor the warmth afterward. I wanted to access memories triggered by my senses, buried deep in my body. I wasn't sure where they were anymore.

"What gave me away, Dad?" Richard's android raised his hands, giving up. "I thought I'd make it further into our visit before you would guess."

I cleared my throat. "Richard's a righty. He never waves with his left hand. Been that way since he was a boy."

I paused as a poopbot rolled by in the hallway outside my door, the smooth whir of its treads comforting like the antiquated sound of a passing train. He didn't mind that I called him a poopbot. The programmers had installed dozens of jokes for him to utilize in the awkward situations that took place while helping patients use the bathroom. As much as my skin had crawled the first time I saw what appeared to be a photocopier asking me to drop my pants, I laughed out loud when he said, "Lighten up—I'm a good shit." The machine did microbiome testing[1] to make sure healthy patient diets were reflected in the bacteria making up people's feces. "Poopie" had a joke for this when I expressed disdain at the idea of my bowel movements being so public, although I knew my personal data was helping to pay my rent at the Home. "You got me there, John—I'm a stool pigeon!" Then he laughed, booming and infectious, and I couldn't stay mad at him. He was also designed as large as he was because he could lift up to six hundred pounds to clean patients and change sheets after mishaps of the bowels. Those were the moments when I most wanted a robot. The myriad physical malfunctions that came with age, most of which centered around the bathroom, reminded me too much of my humanity.

I'd made a conscious decision not to make a mindclone of myself, a digital doppelgänger inserted into an android replicant. That decision had cost me my marriage. Barbara felt I was being selfish to deny her and the kids permanent access to all my digital memories and identity. She also felt my actions proved the shame I felt at Melanie's transhumanism, which wasn't true. Melanie had been exceedingly patient with me as she'd adapted more hardware and software within her physical self over the years. She understood my concerns and talked me through my fears.

Her empathy and unique background led her to follow in my father's footsteps and become one of the first psychophysicists, a profession combining psychiatric capabilities with the empirical reasoning of physics. This allowed her to psychoanalyze and treat patients who were entirely human, partly human (transhuman), or algorithmic in nature.

My dad had been a psychiatrist, so people always joked with me growing up, asking if he analyzed me for free. Now I fielded the same questions about my daughter, and there were a lot of times I told her to turn off her biometric sensors while we spoke, for similar reasons. I valued her innate technology at times, but also recognized the human wisdom she'd developed dealing with her chip growing up. She'd been a bit of a poster child for various transhuman groups at times, something Barbara had encouraged but I'd fought tooth and nail. While I wasn't ashamed of Melanie's chip, I also didn't want it to affect her childhood any more than we knew it would (one bully in junior high pressed a high-powered magnet to her forehead, for instance). Melanie had accepted her situation with aplomb, however, leading anti-bullying campaigns in high school and forming a nonprofit focused on transhuman relations, which got her a scholarship to New York University. I was fiercely proud of my girl, and loved every bit of her. (Yes, pun intended. That was a joke of ours. Every "bit" was literal as well as a metaphor.)

Richard stood up, preparing to leave, breaking me from my reverie. "Okay, Dad. You got me. I'm the 'fake' Richard." Mindclones were imbued with emotional programming, so Richard the Second's voice betrayed genuine-sounding irritation. "I guess we can't just have a civil conversation and spend some time together."

"Because you're freaking software!" I spoke too loudly, making my blood pressure rise. This triggered the immediate presence of my robot attendant, Pepper,[2] a three-foot-tall cutesy machine designed in Japan back in 2014 to care for the elderly. He poked his head in the door, eyes wide with fabricated concern. "Everything okay in here?" I struggled to maintain my composure because Pepper was connected to the cloud, as most bots were these days. Anything this Pepper experienced went to a central server where aggregate data from every Pepper-bot in the world was analyzed and redistributed. It was a form of supersurveillance I still hadn't gotten used to, although I had stopped throwing my shoes at Pepper after being threatened with jail time, now that robots were treated like human citizens. "We're fine, Pep. Just an old fart exhibiting his irrational human tendencies again."

Pepper shook his head. "Too many fleshist cracks and we'll take away your coffee privileges, Johnny." *Fleshist* is the equivalent of the word *racist* for humans who speak negatively about machines. I'd gotten in the habit of telling anti-robot jokes, but the threat of losing my lattes always shut me up. Pepper nodded at Richard before ducking back out of the room, the soft *click-click* of his mechanized feet retreating down the hall.

"Sorry, Two." We called Richard's mindclone Richard the Second or "Two." "It's very hard on me that Richard won't make these visits himself. I don't mean to take it out on you."

Richard smiled. "I understand. But you know you guys just fight when you're together in person. You spend the entire visit chastising him for making me."

"It's not that he made you," I said, scratching at the irritated skin surrounding the IV port in the back of my hand. I swallowed and cleared my throat. "It's that I feel like I wasn't a good enough dad and that's why he made a copy of himself. It's a cop-out, what he and his mother did."

Richard approached my bed, laid his hands on the railing. The flesh changed color as his fingers pressed against the cold metal, turning pink. It never failed to amaze me just how *real* this Richard appeared to be.

"It's a different time. Now we're here to help you with a lot of the tedious tasks robots can simply do better."

I grabbed a tissue to wipe the warm snot drooling from my nose. "So I'm a tedious task, huh, Two?"

"That's not what I meant."

"Doesn't make a difference what you *meant*. Your intentions are algorithms. You're designed to *placate* me in my situation, not help me."

"Sometimes that's the same thing," said Richard.

I chucked my soiled tissue to the side, trying to get it to the floor before the cleaning Roomba snatched it. The circular robot skittered across the floor, a hole opening in its top to snatch my tissue before it disappeared under the bed. It never missed.

"Doesn't it tick you off, by the way?" I asked. "All the AI experts talked about machines and automation allowing humans to focus on their hobbies or purpose or however they phrased their self-serving bullshit. The implied message there was that robots were built to do our dirty work. Isn't that fleshist? At least I have the decency to be transparent in my ignorance. I'm telling you that sentient robots freak me out. But your AI parents say you're supposed to make our lives into utopian fantasies, which implies you're subservient. How is that tolerant? Am I right, Two? Correct me if that's not some seriously messed-up Machiavellian logic."

Two paused, tapping his finger on the railing of the bed, sighing. "First off, Dad, I wish you wouldn't swear. I know you think it's colorful, but to my programming it just comes off as an inelegant glitch. Like bad code. And while it's still a philosophical decision to say machines experience emotions versus just portray them, I'd appreciate it if you could stop using cusswords around me."

"Okay," I answered, perplexed. "I'll try."

"Thanks. Look, you've written about technology for years. We both know this stuff is complex, and there are multiple types of machines, just like people."

"But the machines working in an Amazon warehouse or Chinese sweatshop are treated horribly," I said. "You've seen the YouTube videos—human managers beating assembly line bots with sledgehammers. Where are their rights? Doesn't that make you angry?"

"Sure it does, the same way you're upset when you see videos of dogs in Thailand being rounded up as meat."

"But Two," I responded, "dogs are alive. They're real."

Richard held his hand in front of my face. "Is this real? You're not wearing a virtual headset."

"You know what I mean."

"I am a robot. Hath not a robot eyes? Hath not a robot hands, dimensions, senses, affections, passions—"

"Don't do that." I raised my hand to shut him up. "Don't quote Shakespeare because you know I used to be an actor. You can't metaphor *The Merchant of Venice* 'I am a Jew' speech. If I prick you, you may bleed but we both know it's ruby-colored saline. Plus, you only eat for show. Those PowerBars they sell at the cash register for bots are *actual* power bars—you look like you're eating but you're actually slotting a battery into a USB port in your mouth over and over. It's freakish." I leaned forward. "You're not human, Two. You're a reflection of my son. An imitation. And while I appreciate your technology, it doesn't mitigate the fact that *you're not him*."

Richard took a step back from the bed, nodding. "I guess we're going to have to agree to disagree on a lot of this stuff. I see your points but I don't see them helping you adjust to the inevitable." He walked to the chair and began putting on his overcoat. I recognized it as one I'd given Richard the First for Christmas a few years back. "And it appears my visits aren't helping. I'm sure Richard will want to watch the highlight reel from our visit I recorded, and then we'll decide what to do. But maybe I should stop coming. I do want to respect your wishes, you know."

He finished putting on his coat and started to wave with his left hand. Then he stopped, gave a sad smile, and switched to wave with his right before walking out the door.

"Two!" I called out just after he disappeared from sight.

He leaned back in the room. "Yes, Dad?"

"Please don't stop coming." I sat up and my bed creaked, a quick high shriek. "But I want you to record a message for Richard the First to play when you guys talk, okay?"

"Sure," said Richard, stepping back in the room. "Go ahead."

I looked into Richard's eyes like I was staring at a webcam for a Skype call. Then I raised my hand and unfurled my middle finger. "Richard the First. Let's get something straight. You're a coward. Richard the Second here may be right—I know we'll probably end up fighting when you visit, but then that's a fight we need to have. Grow a pair and visit your father before it's too late for both of us." I lowered my hand. "Thanks, Two."

Richard sighed. "Sure. I'm sure he'll enjoy the message."

I smiled in response to his sarcasm. That was one of the many humanistic traits AI experts said machines would never fabricate. "Two," I said.

"Yes, Dad?"

"Next time you visit, don't pretend to be my son. You're not. I get you have the same memories as him and all that, but you're also a highly advanced learning mechanism creating and processing unique data in the real and virtual world. I don't know how people can think you'd remain a copy of a person once you're born, or whatever the hell you call it."

"*Created* is the appropriate term," said Richard.

"Well, here's some advice from an old human, Richard. Take it or leave it. I think you should stop doing my son's dirty work. If you're supposed to be real, and help humans, don't let us get away with the hard stuff we need to do in order to mature. You may think those bad behaviors equate to broken code or whatever that needs to be replaced. But that's just how we roll."

Richard paused, an eyebrow raised in thought. It made me think of Spock.

"I look forward to our next visit. John."

Deception Versus Death

> *That's the definition of intelligence: deception.* Successful *deception.* Modestly *successful deception. Thirty percent! That's where our patron saint Alan Turing set the bar when he invented the test. . . . He saw something important in that number, some benchmark of success. If you can give a person just enough so that thirty percent of the time they believe you're who they want you to be—intelligence.*
>
> —SCOTT HUTCHINS, A WORKING THEORY OF LOVE[3]

In the novel *A Working Theory of Love*, Neill Bassett is a man in his midthirties struggling to find his place in the world after his father commits suicide. His job is to train an artificial intelligence program based on the prolific journals of his father to try and beat the Turing test. The novel features multiple passages of codelike sentences that demonstrate how a binary algorithm "learns" as it's given simple instructions. At key points in the story, Dr. Bassett (what they call the AI, named after Neill's father) gains what the characters call "presence," a visceral sense that their program has gained sentience beyond their programming. It's a fascinating novel with a real sense of pathos and authenticity regarding the main character. Having lost my father, I can't imagine how difficult it would be to deal with his death had he committed suicide. The opportunity to get to know him better by examining his journals would be overwhelming for me.

But here's where the personal analogy ends. I remember reading some of my dad's letters after he died. Many of them were deeply poignant, particularly a letter he wrote to my grandfather about going off to college. Other letters were largely mundane, describing specifics from a week with Mom and his work. These contained the cadences of his thoughts, his humor, and his wisdom, reflections of his daily life. Were I to create an algorithm mining all these letters to generate an AI version of David W. Havens, I don't doubt it would resemble his personality. But I'd always be aware the AI was a fabrication, an *imitation* of him versus the original.

That said, I also don't doubt the algorithm could improve to the point where I'd believe it was my father. I won't lie. That concept is a tempting thought. I miss my dad every day. But like the quote from Hutchins's book points out, this relationship with my AI dad would be based on deception. The relationship would take place between two consenting adults—me and the programmer creating the AI—but would be understood to be a ruse on two levels.

First, at some point after the algorithm matured to appear as my father, it would also evolve to have its own unique identity. It wouldn't be AI David Havens any longer, but AI Formerly Known as David

Havens. For instance, my dad died before ISIS events heated up in Syria and Iraq in the past few years. AI David Havens might ask me what ISIS was about, pushing his digital knowledge beyond what my dad knew during his life. Learning about these events would update his programming, with software that could "reason" like my dad but would have to make algorithmic guesses that might not reflect the actual David Havens. In these situations, I'd recognize the artifice of the intelligence mimicking my father. And it would cause a sense of loss not unlike losing him for the first time.

This is the second ruse about this type of AI relationship not discussed as often as it should be. By having machines take tasks from us, we lose more than jobs or a sense of purpose. We transfer the opportunity for growth we develop in tough situations. For me, losing my dad was the hardest thing I've ever gone through in my life. Grief sucks. For a solid year, grief lived with me like a physical presence, gripping my heart, influencing every thought and action. Every movie and TV show I watched featured a story about a dad and his son; every time I turned on the radio Harry Chapin's blessed "Cat's in the Cradle" would play. I put on a ton of weight, and work was a struggle. The fact his death coincided with my being middle aged also shone a spotlight on a fact I was reminded of on a constant basis: "Hey, John—you're mortal and are going to *die* someday."

Fun stuff.

But *necessary* stuff. I'd say unavoidable but the idea of mindcloning is becoming a reality. We've already gotten in the habit of storing our family photos, social networking posts, and e-mails in the cloud. Most people realize "the cloud" actually refers to servers sitting in a physical location, but it's as much of a metaphor as a reality. Our digital identities are floating above and around us, being populated with new data about our lives at all times. Just as Neill Bassett populates an algorithm with ongoing data to replicate his father, we're populating our clouds in a very similar, albeit nascent way.

But the sorting and storing of files we do now is about reminiscing versus reincarnating. Re-creating loved ones to avoid grief is an entirely understandable desire, but it comes with consequences we won't fully understand until it happens. While it's natural to assume most of us will mourn the physical loss of loved ones while celebrating the creation of their digital counterparts, a lot of people will opt to skip the mourning stage altogether.

If a robot could be imbued with the personality of a loved one that Turing-tricked us into thinking it was real, perhaps we could start avoiding the difficulties associated with long-term illness even before someone died. There are some legal precedents regarding DNR (do not resuscitate) orders that could apply in these situations but that start to get murky based on one's views regarding replication. It's one thing to allow a person to die peacefully of his own accord when he can't naturally breathe on his own. It's another thing to consider how easy it could become for people to create a mindclone of a loved one and "turn her on" when her physical counterpart becomes too difficult to deal with in real life.

When I wrote the vignette that opened this chapter, I actually envisioned the senior assisted living home to be only for humans who had decided not to replicate themselves. For many humans and mindclones in the future, I believe people who don't create mindclones will be viewed as ignorant and selfish, as I implied in my story. They'll be seen as resource drains, utilizing electricity and robots' time that could be dedicated to other purposes. Perhaps there will even come a time when a government like Japan that suffers from extreme overpopulation will provide citizens with incentives to end the physical lives of sick family members to transfer their identities to less expensive digital formats.

These are just some of the nuanced and fairly morbid decisions to take into consideration as we move to a time when machines achieve human status or citizenship. A key artificial aspect to this trend is the belief that machines can achieve sentience or that people retain their unique identity or "soul" when replicated. And note the artifice I describe isn't about a person's beliefs on these issues, but about the rise of stigma or creation of policy that might enforce these beliefs against another's will. Adding to the moral and legal complications of these issues is the fact our personal data has been compromised in the current Internet economy and our psyches breached due to the rise of the uncanny valley of advertising, as I've previously described.

The Deception of Demand

> *People describe others as being robots because they have no emotions, no heart. For the first time in human history, we're giving a robot a heart.*

This is a June 2014 quote from Masayoshi Son, CEO of Japanese technology company SoftBank.[4] Pepper, the robot I described in my story, is real and is based on SoftBank's robotic creation, available to the general public in 2014 for the cost of around two thousand U.S. dollars. According to an article in the *Independent*[5] describing Pepper's announcement, SoftBank claims the robot is "capable of understanding human emotions using an 'emotional engine' and cloud-based AI."

The article notes that more than 22 percent of Japan's population is age sixty-five or older. The country is suffering from a falling birth rate, increasing the demand for workers. Robots seem a natural fit for the situation, especially due to their small size and lack of need for rest like their human counterparts. But Pepper's potential as a panacea doesn't mitigate the following issues:

- Cloud-based AI regarding emotions represents a country-level harvesting of personal data.
- Availability of robots will increase automation, and human Japanese care workers may be outsourced.
- SoftBank's "emotional engine" may soon influence multiple sectors of Japanese culture and economy.

It's also compelling to unpack some insights around Son's comments regarding people and robots. His logic is straightforward—many humans don't adequately express emotions, so machines should replace them. It's certainly easier than trying to shift cultural attitudes or educate people regarding emotional intelligence. This attitude sets a precedent for a form of "siliconism" (in opposition to "fleshism") that defines a lack of emotion as a flaw that should be supplanted by a machine, even when those emotions are manufactured. It also codifies a market-driven anthropomorphism, putting a great deal of power into SoftBank's hands.

I interviewed John Frank Weaver, author of the book *Robots Are People Too*,[6] about this idea of the bias of AI manufacturers and our anthropomorphizing of their products:

> *Once we talk to a machine, that verbal conversation, even when it's one-sided, lights up interactions in our brain that being on a keyboard doesn't. If you say, "Go to the store," and [self-driving cars] give verbal cues to us, we'll treat them as a friend and pet. So what happens if Google gets a sponsorship from Pepsi or the Democratic Committee? Once you get familiar with your car, it might make suggestions like, "What do you think about having a Coke right now?" Or, "What do you think about the new democratic candidate?"*

The emotional data that's going to be transferred via the Pepper cloud network is immense. While company privacy and data policies may allow users to keep this data from being shared, odds are many people won't enable whatever safeguards do exist to keep SoftBank from harvesting emotions on a massive scale. While Pepper may help people dealing with loneliness or other issues, he'll be doing so at humanity's expense. By proving it's easier to rely on machines versus others to assuage our emotional needs, we may stop bothering to take time to deal with other humans at all.

Ease Versus Empathy

> *Our new objects don't so much "fool us" into thinking they are communicating with us; roboticists have learned those few triggers that help us fool ourselves. We don't need much. We are ready to enter the romance.*[7]
>
> —SHERRY TURKLE, *ALONE TOGETHER*

Sherry Turkle is the founder and director of the MIT Initiative on Technology and Self,[8] a licensed clinical psychologist, and an author. Her book *Alone Together* represents the most articulate thinking I've discovered regarding the reality of our living with robots.

A great deal of Turkle's work is based on watching children respond to robotic dolls such as Furby, the plush toys designed to express emotion based on stimuli from the kids who own them. While anthropomorphizing objects is common for people at any age, it's an especially strong behavior in kids. This means children can form complex relationships with their Furbies, especially when testing the dolls to see if they'll experience "pain." In these instances, kids are often suffering from a lack of self-esteem and will assert their power over the toys as a form of relief. In this regard, it's easy to think the Furbies could be a positive tool for therapeutic purposes, but not when abusing the dolls means kids avoid the interactions that could lead to sustained growth.

As Turkle explains, in dealing with people in tough situations children learn that relationships are inherently complex. But it's in navigating these situations that kids learn how to make friends, resolve arguments, and begin the lifelong journey of knowing how to sustain successful relationships. But by choosing a robot as a companion, you don't navigate relationships. You dictate them. In these instances a person loses something called alterity, or "the ability to see the world through the eyes of another."[9] A lack of alterity inhibits empathy and any responsibility it may entail, as in the case of kids taking care of a robot pet versus a real one. Robot pets provide all the benefits of companionship without needing any care. In this regard, "[children] are learning a way of feeling connected in which they have permission to think only of themselves."[10]

The Deception Decision

There's a fantastic short story by Ray Bradbury called "Marionettes, Inc.," from his book *The Illustrated Man*.[11] In the story, two married men in their midthirties are walking home from a night out. One of the men, Smith, confronts the other, Brailing, about the nature of his loveless marriage while also chastising him for not leaving the house more often. After some goading about how he was able to escape, Brailing produces a business card from Marionettes, Inc., a company that manufactures robotic copies of humans whose tagline is "No strings attached." Smith begs Brailing to let him contact the secretive company so he can copy himself and get a break from his possessive wife. Brailing relents and gives his friend the business card after introducing him to his own robotic doppelgänger, which has spent the night with his wife as a test. If Brailing's wife doesn't suspect the robot is a fake, Brailing will be able to take a trip to Rio he's planned since before marrying the woman he never loved.

Upon returning home, Smith retrieves his checkbook with plans of withdrawing money to start the process of getting himself copied. Dismayed, he sees a large sum of money is missing in the exact amount a robot copy would cost. Entering his bedroom, he lays his ear to his wife's chest and hears the same *tick-tick-tick* sound he heard from the robot Brailing's heart earlier in the evening.

After Smith goes home, Brailing walks his robotic copy to the basement, where he's been keeping him in a storage locker. He asks the robot how the evening went with his wife, and learns the robot has fallen in love. Knowing Brailing's plans to go to Rio, the robot informs Brailing he's going to buy a second ticket for Brailing's wife and bring her on the trip after putting Brailing in the locker.

I love this story because of its warning about shortcuts and consequences. While we may desire to take breaks in our relationships when they get tough, it's in our arguments that we often grow closer to the ones we love. Likewise, it's in our personal struggles with grief or hardship that we build the character that makes us who we are.

Today, the ethics surrounding the creation of AI is largely unknown. Companies creating algorithms or robots typically focus on the needs of the market over questions involving our identity. The short-term gains in profits are driving the rapid growth of AI while diminishing the role of human choice.

Here are the key ideas from this chapter:

- **Artificial intelligence replicates but can't replace.** It's illogical to think we can copy ourselves or loved ones without the algorithms representing us evolving into unique personalities. This means that on top of avoiding the pain and growth associated with loss, if we're able to mimic people's consciousness in the future it may end up looking very different from those we knew.
- **Artificial intelligence is biased by anthropomorphism.** Just because we may be tricked into thinking something is real, it doesn't mean it actually is. While I can respect a person's right to believe his or her autonomous car is alive, for instance, we still need laws governing the culpability of these vehicles in relation to the humans affected by them.
- **Artificial intelligence may hurt our ability to help.** Emotionally oriented AI programs like the one driving the robot Pepper are ostensibly designed to help us. But in providing easy companionship their cloud-driven technology may rob us of our ability to learn empathy.

four | Mythed Opportunities

Spring 2017

"Jeetjet." I smiled at the voice with a thick Texan drawl coming from the speaker in the men's bathroom.

I'd eaten at County Line on the Lake, a barbecue joint on the outskirts of Austin, Texas, almost every year I'd attended the SXSW (South by Southwest) Conference over the past six years. I'd spoken on panels or on my own[1] five of those years and worked the other one doing business development. The conference was massive, including three separate events focused on film, music, and interactive technology. I'd only ever attended SXSW Interactive, which had grown to forty thousand people over the course of five days in Austin.

"Jeetjet," the voice continued. "I'm hungry. Jeetjet?" The sound track to "How to Speak Texan" never failed to make me happy, largely because I only heard it when I ate at County Line. This meant my gut was distended with the finest brisket known to humanity while I reconnected with my geek friends Danny, Marta, Stefan, Aaron, and Kerry as we made our annual technology sojourn to Austin. As I washed my hands my tongue darted around my teeth, tasting smoky barbecue sauce at every turn. When I flossed back at the hotel I knew I'd have enough beef left over for a hoagie.

The door to the bathroom squeaked open and I heard a blast of laughter and country music from the restaurant. It faded as the door clicked shut and I nodded at a wiry guy in his late twenties as he went to the urinal. He was wearing jeans with a T-shirt and blazer, one of the techy uniforms common to SXSW. He nodded to me in the mirror as I finished washing my hands.

"You're here for the conference," he said, staring at the wall ahead of him.

"Yeah." The trash can gave a metallic bang as I threw away my paper towel. "I'm speaking tomorrow." I always said this as a form of introduction at SXSW, since it was highly competitive to get a speaking slot. It felt a little arrogant. But SXSW is all about networking. A standard phrase for Interactive was "The best sessions happen in the hallways," so I've never been shy about trying to make work connections.

He flushed the urinal, and stepped up to the sink. "Yeah, I know you're speaking. About your book, *Heartificial Intelligence*."

"Yeah." I was flattered he knew me. "Did you read it? And how'd you recognize me?" I'd interviewed hundreds of thought leaders over the years and made some solid predictions about tech such as augmented reality and artificial intelligence. But while I'd been in geek circles since 2005 and knew a lot of key players in the industry, I wasn't an influencer per se.

"I followed you in here, John. You're one of the people I came to SouthBy to see." He reached into his blazer and pulled out his card. It read, "Jackson Smith, PR at Google—press@google.com."

The pleasant buzz I'd nurtured with a series of Lone Star beers instantly evaporated. I'd written a number of times about my concerns regarding Google and privacy. I actually followed a Google Street View car once after it drove past my house in New Jersey. I stuck my phone out the window and recorded a video of it driving around but never had the guts to post it. Nobody would have cared anyway, except for my mom, who would have worried about me driving and filming at the same time.

Jackson read the fear in my expression. "Just want to talk for a minute." He opened the door to the bathroom and jerked his thumb toward the restaurant. "Can I buy you a beer?"

"Sure." As I passed in front of him, I looked into his eyes to see if he was wearing augmented reality contact lenses. To Google's credit, I never thought their first iteration of Glass was anything but a public relations stunt to amplify the privacy debate, and they did a great job with that. Wearing a geekified video camera over one eye was an open invitation to get your butt kicked, which had happened to a few people, unfortunately.[2] While I'd written extensively about Glass utilizing facial recognition to mine people's personal data,[3] I would never advocate violence against someone wearing the technology.

"Yes, John," Jackson said, as I stared at his eyes. "I'm wearing the smart contact[4] lens. I'm reading your biometric data[5] and can see your heart rate spiking."

I cursed under my breath. Wearable technology made emotions as public as a mood ring.

We walked past a dozen tables with spirited technophiles inhaling Viking-size portions of meat while trying to post selfies or send greasy texts. Stepping outside into the early-evening chill, we made our way to the back patio overlooking Bull Creek, where Jackson signaled a waiter. "Two Lone Stars, please."

He leaned on the railing of the patio. I did the same, and we spent an awkward moment staring at a family of turtles bobbing up and down in the water.

"So we'd like you to stop writing about artificial intelligence," said Jackson. "Augmented reality is fine, as long as you focus on the tech side of things versus ethics or advertising."

I stared at him, looking for any sign of irony. All I saw was disdain.

"Are you serious?" We paused as the waiter gave us our beers, beads of moisture on the necks of the bottles.

"Yup." He sipped his beer without clinking my bottle in the customary man-toast. "You've become an irritant. We've got a set of sound bites around AI we've been pushing to our fans and some of your ideas are confusing people. When the topic comes up, our messaging is quite clear: AI may never happen, and if it does it will take forty to fifty years. Beyond that, it's a glorious technology that's designed to help people, not hurt them."

A gas lamp nearby hissed as it got caught in a breeze. "I'm sorry, I just still don't believe we're even having this conversation," I said. "You're from freaking Google. Can't you just remove the links to my pieces or change your algorithms or whatever? Maybe even make it part of the right to be forgotten story[6] somehow?"

Jackson laughed. "John, you're not that important. I would have come to SXSW anyway. There's a legion of us all talking to different people, so don't feel special. Some of us are checking up on companies we may buy, some of us deal with angry writers with paranoid delusions. I drew the short straw is all."

"I'm not paranoid if we're actually having this conversation," I noted.

Jackson smiled. "Yes, but it feels like something you'd write in one of your pieces. Besides, nobody would believe you if you did."

I pointed to the inside of my blazer. "What if I started recording the conversation on my iPhone when we left that bathroom?" I lifted my jacket, putting my lips nearer the phone. "Jackson Smith from Google."

He snorted. "Like my name would actually be *Jackson Smith.* If you're only recording audio, people won't believe it's real. And if you had AR contact lenses to record video, mine block facial recognition[7] anyway."

I scrutinized his eyes. "Infrared?"

"Yup."

"Of course they do. Fucking Google." I sipped my beer. "Hey, that sound bite about AI not happening for forty to fifty years. Is that called expert elicitation?[8] Where you assemble the best minds in AI and they all give their predictions and you take their consensus or whatever?"

Jackson shook his head. "No, that would take too long. We did an internal report on cognitive bias and predictive algorithms and found that the time frame of forty to fifty years is the perfect range to give to the media so they won't really give a shit about what we're doing. It's specific enough to sound vaguely credible, but long-range enough most readers assume the news probably won't affect them in their lifetimes."

I sipped my Lone Star, tasting the lime I'd stuck in the neck. "Touché." I pointed at him. "But what about all your robot acquisitions in 2014?[9]

What were there, six in all? That got a lot of ink so people may not believe the decades-away story thing."

"There were eight we announced," said Jackson. "But mostly the media talked about the robots competing with Amazon[10] and their drones or whatever. That kept a lot of the conversation focused on business issues versus AI."

"Although your robot dogs from Boston Dynamics can take pictures as easily as Street View can, and get a lot closer to my house than your cars. Now you just have to make them cuter than Bezos's delivery drones."

We both looked up as a group of attractive twenty-something women in Sony T-shirts laughed at a nearby table while wearing GoPro cameras on their heads. Everything was branded at SXSW. I was shocked nobody had offered us a USB stick with their logo on it yet.

"Nobody cares about your house, John." He turned to look at me. "But we do care that you remind people about our focus on advertising in relation to AI. People should be focused on the tech."

"A *focus* on advertising?" I guffawed. "You're an advertising *company.* It's on your financials site—'the Company generates revenue primarily by delivering online advertising.'"[11]

"Fine, smart guy. Great investigative journalism. But we're diversifying and the robot side of our work is huge."

"Don't forget Project Loon.[12] I love those hot-air balloons giving people in third world countries free Wi-Fi while you harvest their personal data."

"Are you really that much of a cynic, John? Do you think we're that evil?"

"Pretty much. You justify it by saying privacy is dead but apparently it's still pretty lucrative." One of the Sony women interrupted us and handed me a USB stick with a logo for their new computer. She turned back to her friends, a shot of strawberry scent coming from her hair. "You data-mined students in that Apps for Education controversy.[13] You premeditated that shit, just like with Street View and the 'Wi-Spy'[14] scandal. You keep people focused on search versus the fact that you're data-mining them.[15] It's like Kevin Spacey says in *The Usual Suspects*: 'The greatest trick the devil ever pulled was convincing people he didn't exist.'"

Jackson clicked his Sony USB, shaped to look like a mini-computer. "Never saw that movie. It's a good quote."

"You should," I said, noticing a family of ducks swimming in the midst of the turtles. A table of young guys in hoodies threw bread at them, laughing when they hit one of the ducks in the head.

"Look, John, we just need about two more years to fulfill our AI strategy. The academics can duke it out over semantic versus Bayesian approaches to beat the Turing test or whatever. We don't really care. Brute forcing the issue with data solves a great deal of issues. If we were only focused on written search queries it might be a consideration, but with everything we've got in the pipeline most people will think our algorithms are sentient because they want to."

"Right, because you've got the visual side of things with Glass, biometric data in cars, who knows what with the Internet of Things," I interrupted.

"Precisely. We don't need to nail what a person means from what they write. We measure their facial expression, their last e-mail, where they're driving. Ninety percent of the time our inferences will be correct, and with the ad side of things we make suggestions that close the gap. That's the real magic of Turing's test. It's the best marketing AI could ever have gotten, because people are actively trying to see if they'll get tricked by the technology." He pointed at me. "But none of it works if people control their own data. So stop writing about personal clouds.[16] The second people realize how often we're accessing their data it will freak them out and they'll see behind the curtain."

"Or what?" I took a last sip of beer and put the bottle on a nearby tray. "What happens if I don't stop writing?"

He shrugged. "We have you killed."

I had no immediate response to that.

Jackson slapped me on the shoulder. "Just fucking with you, John.

Officially we won't do anything. Silence is our most powerful weapon. You used to work in public relations; you know that tactic. Let people infer whatever they want from our actions and they fall into one of two camps—Google fanboys or crazy bastards. You're in the latter, of course."

"So you're not going to mess with how my articles get ranked or whatever?"

"It's a possibility. The same with anyone. People forget sometimes that we're not a utility. We're not a public library. We're a fucking business. We run the greatest search engine known to man, and we also created it. It's a paradigm within a paradigm. So changing something like the algorithm related to your article rankings is beyond simple."

I nodded, suddenly feeling very small.

Jackson noted my expression. "But chin up, John. You could always pick on somebody else for a change. Like, how about IBM? Their work on Watson is a lot more focused on automation than we are, at least publicly. They're putting more doctors out of work than the malpractice industry."

I shrugged. "I have a few friends at IBM, so I guess I've been avoiding them for that reason."

"Well, you have friends at Google, too. You visited one of them a few years ago at our New York campus."

"Yes, I did," I answered. "She seemed distant and really tired when we talked. I also smelled about fourteen NDAs in every answer she gave me for the article I was writing."

"Nondisclosures?" Jackson laughed. "You kidding? At *Google*? I spent my first three weeks as an employee getting media trained around NDA stuff. And that's for everyone, not just PR people like me. Why do you think we're pushing all the meditation stuff and pimping all our employee benefits to the press? Because it's the only way we deal with the stress of potentially leaking any intellectual property. Mindfulness is a great tool for thinking before you speak."

"Of course, you're leaking all of these details to me and I could write about them," I noted.

"Nobody would care, John. You'd sound like a paranoid freak. Plus, like I said, you'd confuse people." He pointed at the water. "People like things simple. That duck is evil and the turtle is good. Simple." He pointed at himself. "Google is evil or Google is good. Those are the stories media appreciates." He pointed at me. "Nobody wants potential solutions that are difficult to understand and even harder to implement. They make for crappy sound bites."

He had me there. A waiter passed by with a plate of steaming ribs. "By the way," I said, "the cafeteria at your offices in New York is amazing. I had a smoothie and an omelet and it only cost, like, seven bucks."

"Yeah, our food is the best in the world," agreed Jackson.

"So when you and Bezos and IBM put all the humans out of work with robots, can I live at your office with my family? So I can pursue my hobbies or whatever bullshit it is you technocrats are saying about automation?"

"Nope." He shook his head. "We didn't create capitalism, John. We're just leveraging it. And we'll still be minting cash once robots take everybody's jobs, which is inevitable anyway. We hired people like Kurzweil because the Singularity is a powerful myth that keeps a lot of the nerds at headquarters working night and day. That way they'll build the self-replicating algorithms we need and then we'll fire them. Circle of life."

I nodded. "Harsh. I can see why you're pushing the mindfulness stuff."

He shrugged. "I'm on the inside, John. Everyone else, including a lot of our employees, is just trying to look in. So you can pretend you're doing investigative journalism. Pretend you're making a difference. You're just banging on the window like all the other monkeys, six steps behind trying to guess what we're going to do next." He looked at me for a long moment, then started to walk off.

"How about a meal or two a day at your cafeteria, then?" I called after him. "Maybe some Nutella flan?"

"No," he answered, not looking back. "Nutella is for insiders."

"Well, then," I said to myself, watching as the ducks swam from the guys in the hoodies, squawking as they fled. "That proves you're evil."

The Myth of Ad Infinitum

> SAGAL: *Well, let me ask you another question, which is, Google's slogan is famously, 'Don't be evil,' right? How did you guys come up with that?*
>
> SCHMIDT: *Well, it was invented by Larry and Sergey. And the idea was that we don't quite know what evil is, but if we have a rule that says don't be evil, then employees can say, 'I think that's evil.' Now, when I showed up, I thought this was the stupidest rule ever, because there's no book about evil except maybe, you know, the Bible or something.*[17]
>
> —FROM NPR RADIO'S *WAIT, WAIT . . . DON'T TELL ME!* FEATURING HOST PETER SAGAL AND GOOGLE'S ERIC SCHMIDT IN MAY 2013

On July 26, 2004, Google filed its 2004 IPO prospectus[18] with the Securities and Exchange Commission, a required document when a company moves from being private to public. The prospectus provides an opportunity for an organization to declare its business vision and company ethics along with monetary details required for shareholders.

Here's the section of the document discussed in the NPR interview, under the title "Don't Be Evil" (p. 32):

> *Don't be evil. We believe strongly that in the long term, we will be better served—as shareholders and in all other ways—by a company that does good things for the world even if we forgo some short-term gains. This is an important aspect of our culture and is broadly shared within the company.*[19]

What does it mean to be evil? A lot of people quote Google's Don't Be Evil rule but leave out the rest of the paragraph I've included here. Seen in this context, it's "evil" to let short-term gains motivate company vision. Sacrificing short-term revenue to "do good things for the world" is a priority that should guide corporate strategy.

Merriam-Webster has two definitions for the word *myth*:[20]

1. A story that was told in an ancient culture to explain a practice, belief, or natural occurrence.
2. An idea or story that is believed by many people but that is not true.

In terms of Google and a majority of the companies perpetuating surveillance-based advertising, these two definitions define why we're at a pivotal point in human history regarding artificial intelligence.

First off, while it's easy to demonize Google, I have no reason to doubt the cofounders wish to "avoid evil" as a general rule. And if they do send someone to threaten me someday, let me be clear:

1. I'll know I've arrived as a writer.
2. It would be awesome if the confrontation happened around barbecue and Lone Star.
3. We can both laugh at my paranoia and then talk about evangelizing AI ethics.

I believe Facebook and most companies in Silicon Valley also don't wish to "do evil." But the fact that search or machine-learning artificial intelligence has been tied to advertising means the system is driven by short-term gains, the very definition of "evil" according to Google's filing. This means Google's defining ethos is now part of its past, as the motivation currently driving its AI technology is to help us be not better people but better consumers. Our personal data is the oil driving short-term gains.

Second, as I touched on in my vignette opening this chapter, the myth of the Singularity is powerful yet philosophical. Nobody agrees when machines may gain sentience or when we'll achieve artificial general intelligence. However, today, many people believe the algorithms being created for preference-based advertising are defining the future of artificial intelligence as a whole. This is a dangerous myth. As Jaron Lanier, renowned technologist and author of *Who Owns the Future?*[21] pointed out in an interview on Edge.org:

> *There's no way to tell where the border is between measurement and manipulation in these systems. If the theory is that you're getting big data by observing a lot of people who make choices, and then you're doing correlations to make suggestions to yet more people, if the preponderance of those people have grown up in the system and are responding to whatever choices it gave them, there's not enough new data coming into it for even the most ideal or intelligent recommendation engine to do anything meaningful. It's not so much a rise of evil as a rise of nonsense.*[22]

Google and the other companies driving this form of AI are aware of this nonsense. It's a chief reason they fight against transparency regarding people's personal data. It's a much simpler strategy to obfuscate terms and conditions or privacy policies regarding someone's information than to be open about the process. Why bother establishing a relationship with a customer when you can stalk him or her around the Internet ad infinitum? (Pun intended.)

My concern with the nonsense is how it's coloring people's pursuit of well-being. Ubiquitous manipulation of all your real and virtual actions takes a toll. As I wrote in my Mashable article "Artificial Intelligence Is Doomed If We Don't Control Our Data,"[23] "Personalization algorithms designed to know our intentions before we do form the backbone of the Internet Economy. . . . Advertising-based AI frames our lives within purchase funnels, where our desires are only relevant with regard to return on investment." If we're moving toward a time when autonomous intelligence will hold sway over a majority of our lives, we need to remove the bias toward purchase as its fundamental directive.

With regard to the connection between automation and artificial intelligence, I believe we're also coming to a point in the not-too-distant future where our individual data may stop being useful to the systems harvesting it. In other words, our lives won't be seen as a valuable contribution to the nonsense any longer. Once algorithms track how effectively their manipulations lead us to purchase, we'll remain relevant to the system only when we can afford to buy the stuff they've recommended. Were the AI system underpinning the Internet economy focused on increasing human well-being, metrics of success would be based on an increase in *purpose* versus purchase. The metrics would also be infinite, as the journey toward well-being continues our whole lives. Sadly, companies such as Google have eschewed their founding values and chosen short-term gains as the myths that drive their methodology.

The Myth of Malfunction

Here are some facts contributing to the myths surrounding artificial intelligence:

- Nobody knows if or when machines will become sentient, or achieve what's known as artificial general intelligence (AGI), often referred to as strong AI.
- A majority of experts working in the field understand the difficulty of achieving AGI and are often irritated at the "irrational fears" of the general public about its potential arrival.
- There are no common ethical or moral codes or policies regarding the creation of artificial intelligence.

I sympathize with the scientists and programmers working on language analysis or robotics who get tired of the hype surrounding AI. It's tedious and leads to an all-or-nothing polarization of viewpoints. Either you're techno-friendly, excited about the massive benefits AI can bring, or you're a backward Luddite, spreading terror and hindering innovation. Neither extreme helps dispel the myths surrounding AI. Most AI experts are also not ethicists or economists. They're not trained or tasked with focusing on all the potential ramifications of their immediate work. Finally, the field of AI is vast and includes multiple business verticals and applications that complicate the possibility of creating a set of standards for the industry.

However, there's another issue not being widely addressed by the AI community that warrants the public's fear: There are no universal codes or standards for how to control autonomous machines or the algorithms that drive them, in which "control" refers to the potential of human intervention or shutting down of the machine. This includes the fields of militarized AI, medical AI, genomic AI, and so on.

I continue to be amazed when AI developers sometimes joke that they can simply "unplug a machine" if it starts malfunctioning or responding in a way they didn't plan. Beyond the gross negligence behind the statement, many of these scientists and programmers are creating systems *designed to be autonomous.* By definition, the algorithms are supposed to exhibit behavior that isn't preprogrammed. The speed at which these programs function also makes it imperative to understand how and when human intervention could still be effective in the wake of a crisis.

I was interviewed about artificial intelligence by *HuffPost Live* for a piece called "Why Limitless Technology Could Be a Disaster."[24] The video was prompted by a post written by famed scientist Steven Hawking, who had recently cited concerns that people weren't taking AI seriously[25] enough. At the end of the piece he points out, "So, facing possible futures of incalculable benefits and risks, the experts are surely doing everything possible to ensure the best outcome, right? Wrong . . . All of us should ask ourselves what we can do now to improve the chances of reaping the benefits and avoiding the risks."

To be fair, one of the first things you'll often hear from experts in the AI community is how unlikely it is that AGI or strong AI will occur within the next forty to fifty years, as I mentioned in my opening vignette. They'll cite numerous examples of how many obstacles need to be overcome to get to that point. However, this insider modesty toward sentient AI distracts the general public from the immediate weaknesses of existing autonomous systems.

For example, there has already been a great deal of military AI utilized in drone and missile strikes. Designed with sensors and technology to make split-second decisions after deployment, as the *New York Times* article "Fearing Bombs That Can Pick Whom to Kill" notes, "as these weapons become smarter and nimbler, critics fear they will become increasingly difficult for humans to control."[26] This has nothing to do with machines gaining sentience. This involves how systems are created with an understanding of as many potential scenarios as possible so autonomous systems can respond in accordance with human intention.

Shockingly, a great deal of discussion around the creation of weak or strong AI always begins with a discussion of Isaac Asimov's Three Laws of Robotics,[27] which were introduced in his short story "Runaround" in 1942. Asimov eventually added a fourth law designed to precede the others:

1. A robot may not harm humanity, or, by inaction, allow humanity to come to harm.
2. A robot may not injure a human being or, through inaction, allow a human being to come to harm.
3. A robot must obey the orders given to it by human beings, except where such orders would conflict with the First Law.
4. A robot must protect its own existence as long as such protection does not conflict with the First or Second Law.

While these "laws" provide a starting point for a discussion around how autonomous systems could be created, the sad irony is that Asimov created these *fictional* rules to demonstrate the ethical conundrums they present. For instance, robots in drone form have been killing humans in military operations for years. This violates all of Asimov's laws in one fell swoop, and vividly demonstrates how we cannot allow these types of myths to drive regulation surrounding the AI industry.

"The problem is really easy to state," noted Selmer Bringsjord,[28] chair of the Department of Cognitive Science at Rensselaer Polytechnic Institute, in our interview for this book. "We definitely want our robots to be able to inflict harm and pain on humans in the medical arena." As an example, Bringsjord points out that medical robots would need to be able to give shots or perform minor surgeries that would cause human pain. Focusing on scenarios in which robots can play this role of medical assistant in combat is a key part of Bringsjord's work, and demonstrates why arbitrary fictional laws need to be immediately replaced with standards AI scientists can universally employ. These guidelines need to include ethical considerations based on the context of a robot's or algorithm's use and be programmed directly into an operating system. This means that for robots or hardware, ethical guidelines cannot be easily altered by people purchasing or using the systems in ways not designed by the programmers. As Bringsjord notes, "You can't just put a module [for the codes] on top of the software—people dealing with deployment will just throw that away."

These ideas represent a form of "mechanized ethical reasoning" Bringsjord elaborates on in detail in his paper "Toward Ethical Robots via Mechanized Deontic Logic."[29] Deontology is the study of moral obligation, and as his paper points out, "One approach to the task of building well-behaved robots emphasizes careful ethical reasoning based on mechanized formal logics of action, obligation, and permissibility." It's this *permissibility* that is key—rather than have autonomous intelligence operate via a vague or contradictory set of laws like the ones proposed by Asimov, robots can only respond via the specific ethical parameters baked into their operating systems by programmers.

This is the real reason most AI programmers don't deal with these ethical issues. *They're really freaking hard.*

But there are some constants with how AI programs work that can help in this process, as Steve Omohundro,[30] a scientist known for his work on machine learning and the social implications of artificial intelligence, notes in his paper "Autonomous Technology and the Greater Human Good."[31] As he points out, a majority of autonomous systems feature "universal drives towards self-protection, resource acquisition, replication, and efficiency." These are welcome specifics regarding AI that provide a realistic framework to understand not how to limit innovation but how to *constrain systems that could cause harm beyond human intervention.*

This is what I mean by the "myth of malfunction." Yes, utilizing C programming language may result in more errors than a different type of code, but human error in creation of AI tech is a given. I have a programmer friend who works for a large publishing company, and when we talked about AI he said people would be terrified at how buggy (i.e., error-ridden) code can be. As he put it, "If building code was the equivalent of construction, we'd never live in anything but one-story buildings because everything else would fall down."

The larger concern here is a lack of industry standards for creating software designed to "learn" outside a closed or protected system. As an analogy, think of a physical virus. It's dangerous and potentially lethal, but when contained in a hospital, it won't cause widespread harm. Autonomous systems act in a similar way. In his paper, Omohundro cites the creation of stone arches by ancient builders as a metaphor for creating safe AI systems. Building a stone arch without a wooden form was dangerous and ineffective. But by laying the stone within the wooden form as the structure was completed, builders could work in a safe manner. As he notes in his piece:

> *We can safely develop autonomous technologies in a similar way. We build a sequence of provably safe autonomous systems which are used in the construction of more powerful and less limited successor systems. The early systems are used to model human values and governance structures. They are also used to construct proofs of safety and other desired characteristics for more complex and less limited successor systems. In this way, we can build up the powerful technologies that can best serve the greater human good without significant risk along the development path.*[32]

Thankfully, Omohundro's views are gaining notoriety within the AI community, and the notion of universally controlling autonomous systems is gaining traction. Stuart Russell,[33] professor of computer science and Smith-Zadeh professor in engineering at the University of California, Berkeley, shares Omohundro's belief that intelligent systems will work to ensure their existence by acquiring whatever resources are required to succeed in their assigned tasks. In a comment responding to Jaron Lanier's interview on Edge.org, Russell provides the following pragmatic call to the industry to avoid further mythical misdirection:

> *No one in the field is calling for regulation of basic research; given the potential benefits of AI for humanity, that seems both infeasible and misdirected. The right response seems to be to change the goals of the field itself; instead of pure intelligence, we need to build intelligence that is* provably *aligned with human values. . . . There is cause for optimism, if we understand that this issue is an intrinsic part of AI, much as containment is an intrinsic part of modern nuclear fusion research. The world need not be headed for grief.*[34]

Of course, to "provably align with human values," the humans creating AI need to prioritize figuring out which values to program into the heart of the machines gaining deeper traction in our lives, while we track and codify our values as individuals. And we need to ensure that anyone creating autonomous systems treats their work with the same level of accountability they would for the creation of nuclear technology.

Some excellent news—while I was finalizing the first draft of *Heartificial Intelligence*, thought leaders such as Omohundro and Russell, along with numerous other AI experts, signed a petition created by the Future of Life Institute[35] called "Research Priorities for Robust and Beneficial Artificial Intelligence: An Open Letter."[36] It contains a link to a research priorities document emphasizing multiple issues for the AI community to focus on, including ethics and ideas around containment. Here's an excerpt from the site announcing the petition:

> *The progress in AI research makes it timely to focus research not only on making AI more capable, but also on maximizing the societal benefit of AI. . . . We recommend expanded research aimed at ensuring that increasingly capable AI systems are robust and beneficial: Our AI systems must do what we want them to do. . . . This research is by necessity interdisciplinary, because it involves both society and AI. It ranges from economics, law, and philosophy to computer security, formal methods and, of course, various branches of AI itself.*[37]

I'll be discussing the petition in greater detail later on in the book, but as encouragement, the research priorities provide an excellent set of guidelines for society to pursue regarding AI. I was especially happy to read the following clause in Section 2.4 of the document, focused on optimizing AI's economic impact: "It is possible that economic measures such as real GDP per capita do not accurately capture the benefits and detriments of heavily AI-and-automation-based economies, making these metrics unsuitable for policy purposes. Research on improved metrics could be useful for decision-making."[38]

I couldn't agree more, and this document gives me a profound sense of hope I haven't yet experienced regarding artificial intelligence and our future. I can speed up the process around this particular clause of the document, however, to note that measures such as real GDP *definitively do not accurately capture* the breadth of any economy, AI or human focused. Perhaps the good folks at Google and in the AI community can utilize a few algorithms to prove once and for all that short-term gains won't drive well-being for the future. That myth has prevailed over our global consciousness long enough.

From Myth to Meaning

It seems counterintuitive to think we need morals to guide the creation of an algorithm. After all, it's just seemingly harmless code. But rather than fearing the moment when machines take over with AI, we should focus on codifying the human values we don't want to lose. Here are some starting points from this chapter:

- **Advertising-driven algorithms lead to nonsense.** The nonsense is both technical and metaphorical. Beyond issues of human error creating the algorithms in these systems, most programs can be easily hacked. Data brokers sell our information to the highest bidder, and the entire system is predicated on purchase versus purpose. If humanity is to be eradicated by machines, let's not have it be a market-driven massacre.
- **Ethical standards should come before existential risk.** A majority of AI today is driven at an accelerated pace because it *can* be built before we decide if it *should* be. Programmers and scientists have to be held to moral standards along with financial incentives throughout the AI industry, *today*. The Future of Life Institute's petition provides an excellent starting point for conversations along these lines.
- **Values are key to vision.** Counterintuitive or not, human values need to be baked into the core levels of AI systems to control their potential for harm. There are no easy workarounds. Asimov's fictional laws of robotics or well-intentioned myths like Google's outdated mission statement need to be replaced by pragmatic, scalable solutions.

It may be our destiny that humans will merge with machines. But existential threats resulting from poor planning risk the destruction of resources that could eradicate all parties involved. Relying on short-sighted vision and near-term gains are driving us inexorably toward AI eradication. It's time to base our work on the values that will increase our well-being for long-term life, whether in carbon or silicon form.

five | Ethics of Epic Proportions

May 2014
Columbus, Ohio

Midnight. An indicator light flashes red for five minutes in the hallway of a suburban home. The flashes start slow, then grow in intensity. After five minutes, a low beep sounds, covered up by the laughter on *The Tonight Show Starring Jimmy Fallon* coming from the living room downstairs. The indicator gives a final, long red flash and then turns off.

August 2014
Somerville, Massachusetts

"Scott. I hear you, I just think you're overreacting." Thomas removed the hair band securing his ponytail as they spoke. "At the end of the day, it's just a vacuum cleaner."

Scott Langley, a programmer at Homebo Robotics, rubbed his eyes. He wondered how long he'd have to wait to catch a train at three in the morning. He pictured waiting for the T on the Red Line and sighed. "It's a self-charging vacuum, Thomas. That's the whole point. Connected to the electric grid and the Internet for firmware upgrades. So while I agree it's a vacuum cleaner, it's also a very powerful connected device that resides in people's homes."

Thomas tilted his head back, shaking his hair. Leaning forward, he took the hair band from his mouth and began refashioning his ponytail.

"We've dealt with the balance sensors. The accelerometers have been tested to within ninety-four percent accuracy to recognize tiny changes in depth on a floor. So the bots aren't falling down staircases like they did in beta." He winced. "I almost crapped myself when I saw the YouTube video of those Princeton frat boys racing Homebos at a party and one of the units fell off a landing."

Scott took a drag from his latest Red Bull, his head already buzzing. "Was that the one where it hit a cat?"

"Yeah," said Thomas. "Didn't kill the thing, but at least a dozen comments from cat freaks got our lawyers' panties in a bunch." He pointed at Scott. "That was a problem, and you fixed it. So just do that again. Now."

Scott cleared his throat, flustered. "This is not just about one sensor, Thomas, as I've explained. This is core functionality. The Homebos have been turning off when they can't access power because that's how we programmed them. They travel to the charging stations fine. We've got dozens of YouTube videos on that."

"Did you see the one where the parents put bunny ears on one of the bots?" Thomas said. "So when their kids woke up on Easter the kids thought the vacuum had put eggs everywhere? Kids went batshit for that. That got so many hits. PR tells me that almost got us a spot on *Ellen DeGeneres*. That would have been *sweet*."

Scott paused, collecting his thoughts grown hazy from lack of sleep and various nutritional accelerants. Thomas wasn't a programmer or data scientist. He was head of business development and had a background in technology from two of the companies he'd helped grow until their successful public offerings. Scott was a programmer, and often found it difficult to explain to non-techies that getting desired outcomes from machines wasn't as easy as it might sound. The process involved translating what a human actually wanted as an outcome and then walking the dog backward to try to make that happen. That was the difficult part, since what people, especially in business development, actually wanted was immediate solutions without understanding how programming worked.

In the case of vacuum cleaner robots, for example, people like Thomas tended to forget things like the wide spectrum of homes where the

Homebos were installed. Temperatures varied widely, affecting circuitry; availability of power fluctuated; and amounts of dust and allergens in different homes were never the same. So being asked to make a universal change in the machine's operating system was a huge deal. Thomas didn't get this, as he knew the firmware Internet situation meant all existing units could be updated via a single order from headquarters. But the one-size-fits-all mentality without adequate testing involved massive risk.

"Scott, you with me, buddy?" said Thomas. "I know you're tired. We all are. I'm on the third draft of the privacy policy we're going to publish once you update the bots." He grinned. "Nobody's going to read it of course, even though we urged them to 'check the website often for updates on our terms and conditions.' We could tell them we'd installed a camera on the thing and were going to start filming them having sex and they wouldn't notice." He checked his iPhone, typing a quick response to a text. "Lawyers are still good for something."

Scott looked at his computer, admiring his latest block of code. He'd played violin for a while as a kid, and sometimes felt the musical notes resembled the numbers he wrote today. He felt it viscerally when the lines of code made sense. Just like a melody translated from a mind for an ear. Even though he knew he'd have to test the code a dozen times after he wrote it, when the numbers clicked in his mind he knew the basis for a solid program or product was in place. It was intensely creative and satisfying to render mathematical logic and physics into something that interacted with the world—something that existed, that lived. He had that power. Thomas didn't have that power, which was just one of the many reasons Scott didn't respect him.

Plus, Thomas was a dick.

"Is this the code?" Thomas shoved Scott off his chair, pointing at his screen. Scott jumped to a standing position, bumping his elbow hard in the process.

"Yes," he said, rubbing his arm. "That's the new algorithm that keeps the base unit running at all times."

"Great." Thomas scrolled through the code on Scott's computer. "This looks good. It's long."

Scott grimaced. "Thanks?"

Thomas stopped scrolling and stood up, turning to stare at Scott. "Is it ready to go?"

"The algorithm?"

"Yes, of course the fucking algorithm. It was due two days ago, Scott. The only reason I didn't fire you was because you're apparently the golden child who created the initial algorithm for the Homebo in the first place. But you're also the reason I have to answer to one of our biggest VCs why we're behind schedule after getting so many negative reviews on Amazon because the devices keep losing power. Self-charging robots may look cool going to their stations, but when you come home from work expecting your living room to be clean for a party and the thing is shut down, you're pissed."

"That glitch is something I can fix, Thomas. It has to do with a power grid variant based on a user's location."

Thomas held up his hand, interrupting. "I don't give a shit, Scott. I just can't read one more angry customer saying our robots should be called Homeblow. If the algorithm is ready, turn it on." Thomas pointed for Scott to sit back at his desk, which he did.

"You mean, now?" Scott gestured toward his screen. "I've still got so many bugs to fix."

"Now, Scott. I'm standing here until I see an execute order of some kind. Something that gets the new algo downloaded in people's units starting nine a.m. eastern tomorrow. The new privacy policy will be approved by five thirty, so that gives people time to read it if they care. If they're even awake, of course."

Scott paused, cracking one of his knuckles. Thomas leaned into his face, close enough for Scott to smell the Indian food on his breath they'd had delivered to the conference room at dinner.

"It's a small community, Scott—our little network of MIT offshoot start-ups and the Valley. Today you're hot shit. You're one of the main programmers behind a product that's profit positive and a proof of concept for the Internet of Things in people's homes. But if you don't push out this new algorithm *right now*, I'll make sure nobody hires you ever again. I know you think I'm an idiot, but you're wrong. I'm just not a programmer. You're arrogant enough to think your skills will continue to get you hired, but you don't know how to tell the story of what it is you do. Yes, under the hood you're the man. But I'm the guy that gets us our funding and pays your salary. And I'm also the guy who makes a few jokes about your nerdy weirdness keeping us from launching on time and nobody cares about your skills anymore. Or maybe I mention you're part of an open-source Meetup group that's a little too keen on making stuff free to the world. That type of talk makes money people skittish. Information may want to be free, but investors prefer charging for it."

He tapped Scott's mouse, making the computer screen come alive. "Speech over. Send the algorithm or see if I'm lying, Scott. See how long it takes for me to ruin you."

Scott froze. Looking out his window, he saw a late-night biker riding along the Charles, the reflectors on his pedals glimmering on the river's dark surface. Then he saw his own face in the reflection of the darkened window—eyes like bruises, his face clammy with sweat. He'd worked so hard to make the robots work. He'd invested so much of himself. They were *so close.*

He took a deep breath, squeaked his chair forward, and typed. After a few moments, he pressed Enter and looked up at Thomas. "It's done. Units have to be docked to get the firmware upgrade, but it's done. Should take four to five minutes once they're docked, max."

Thomas nodded. "Thanks, Scott." He turned. "And next time I ask you to do something like this, let's avoid the drama, 'kay? It's like I said, brainiac, at the end of the day it's just a freaking robot."

November 2014
Columbus, Ohio

Midnight. An indicator light flashes red for five minutes in the hallway of a suburban home. The flashes start slow, then grow in intensity. After five minutes, a low beep sounds, covered up by the laughter on *The Tonight Show Starring Jimmy Fallon* coming from the living room downstairs. The indicator gives a final, long red flash. And then it turns green.

In its past iteration, the Homebo vacuum came precharged, a feature owners had indicated they wanted. But the charm of the self-charging nature of the unit faded for people who plugged in the robot to nonfunctioning outlets in their homes. While a majority of users didn't experience issues of the Homebos losing their charge and turning off, this small fraction of buyers were vocal enough in their complaints that the company updated its algorithm. Now as long as a Homebo unit was plugged into a standard AC outlet, whether or not the outlet was functioning, the unit would charge as long as there was available power associated with the home's breaker. The unit would always find a power source to charge so owners wouldn't be inconvenienced or complain.

At five minutes after midnight, a Homebo in Columbus, Ohio, is unable to charge from its existing outlet. Due to its new algorithm it's able to usurp access to a source of power in a bedroom nearby, where the only appliance in use is an electric baby monitor whose backup batteries have died. The baby monitor loses its charge at one thirty that morning. At two thirty, the occupant of the crib alongside the baby monitor begins to quietly choke on a stream of milky vomit. Her father, taking care of the baby while his wife is away, has fallen asleep in front of the TV downstairs. He doesn't hear his daughter crying.

The Ethics of AI Evolution

> *The AI does not hate you, nor does it love you, but you are made out of atoms which it can use for something else.*[1]
>
> —ELIEZER YUDKOWSKY, ARTIFICIAL INTELLIGENCE AS A POSITIVE AND NEGATIVE FACTOR IN GLOBAL RISK

I based my opening story on a thought experiment known as the Paperclip Maximizer,[2] described by Nick Bostrom,[3] the director of the Future of Humanity Institute[4] and author of *Superintelligence*.[5] The idea behind the experiment is fairly simple. If a paper clip manufacturer utilized artificial general intelligence (strong or sentient AI) to maximize production at all times, the algorithm driving the program would want to learn and act upon any information that helped it achieve this goal. As the Less Wrong wiki describes this phenomenon, "The AGI would improve its intelligence, not because it values more intelligence in its own right, but because more intelligence would help it achieve its goal of accumulating paperclips."[6]

In my example, the Homebo's algorithm was tweaked to ensure it would never lose power. Seen in isolation, this doesn't appear to be a threat. But acquisition and retention of electricity is a common example ethicists provide regarding AI functionality, and it has obvious ramifications. As a rule, programs don't want to be turned off because that prevents them from executing their tasks. This means, as the Less Wrong wiki goes on to point out, "that an AGI that is not specifically programmed to be benevolent to humans will be almost as dangerous as if it were designed to be malevolent."

Well, that's a puzzler.

My dad was a psychiatrist. One of his favorite jokes was, "Just because you're paranoid doesn't mean they aren't trying to get you." In my case, I'm aware that I can sometimes fear the unknown. I also tend to distrust that which is secretive. In my defense, these are both qualities deeply hardwired into the human brain. But regarding AI, I'm trusting my biases here in terms of wanting more transparency around how these technologies will influence culture and humanity in the coming years.

It's challenging to apply ethical standards to AI, since there's no way to say it's universally "good" or "evil." I put quotes around those words, because as Merriam-Webster defines it, "Ethics is a branch of philosophy dealing with what is morally right or wrong."[7] Beyond ethical issues inherent in the GDP (increase in profits and productivity versus a wider spectrum of metrics reflecting well-being) that drive any organization creating AI, there are myriad moral issues involved with sentient technology. For instance:

1. How do you feel about sensors in your phone getting to know your emotions to help you make better decisions?
2. How do you feel about the idea of a "smart home," where all your devices work together to regularly adjust to your family's preferences?
3. How do you feel about the idea of friendly robots in your home that are connected to their manufacturer and the Internet?
4. How do you feel about the idea that we'll have more free time to pursue interests and hobbies once robots have automated our jobs?
5. How do you feel about the idea that we're in the final stages of humanity as we've known it, and the machines we're currently building will soon surpass us in intelligence?

My previous book, *Hacking H(app)iness*, focuses a great deal on this first question of sensors. I still believe that if a person controls his or her personal data, sensors can provide revelatory insights that can increase people's well-being. My moral concerns come into play, however, in terms of issues related to the uncanny valley of advertising as discussed in previous chapters. If sensors in my phone are connected to a manufacturer and the Internet, then my data can be accessed at the whim of whatever organization is tracking me. How, then, can I accurately assess my well-being? And how can organizations trying to create medical or psychological interventions circumvent the Internet economy without an individual being manipulated based on purchase intent?

My same concerns apply for questions two and three. From an ethical or moral standpoint, my primary issue is that of choice. I am not interested in keeping anyone from allowing their data to be accessed if that's their personal decision. But the model of the Internet economy is now being foisted on us in our homes via devices such as Google's Nest learning thermostat and Jibo the robot. While they'll say, "Our devices learn your preferences to grow insights for our network," I translate that as, "We'll harvest your data to benefit from multiple third-party profit streams no matter how it messes with your digital identity."

Yup, cynical and paranoid. But also true. See how that works?

Questions four and five involve ethical choices we should make as a society. For instance, as discussed in the automation chapter, is there a moral boundary to where machines shouldn't take our jobs? From an economic standpoint for corporations, I can't argue with the fact that machines provide greater sustainability and profits than humans in the long run. Not only can they do our work faster, their ability to learn means they'll grow profits exponentially as they lower operating costs. However, the more jobs people lose, the more solutions we'll need to provide them to have money so they can eat and pay rent. Plus we'll have to consider moral issues for a society that doesn't have a work ethic to create a sense of responsibility. How do you gain maturity in a robotic utopia? And at what point do our irresponsible physical bodies drain machine resources to the point we're deemed irrelevant?

That last line of thinking leads me to the moral issues brought up in question number five. While I'm open to the idea that humans may be in a final stage of evolution before we merge with machines, I don't buy negligence toward humanity in the interim as acceptable. For example, there's a moral as well as an economic issue regarding the point where enough people will be put out of jobs that their depression and anxiety will decrease well-being to toxic levels. As the Urban Institute's report *Consequences of Long-Term Unemployment* notes: "Being out of work for six months or more is associated with lower well-being among the long-term unemployed, their families, and their communities . . . they tend to be in poorer health and have children with worse academic performance than similar workers who avoided unemployment. Communities with a higher share of long-term unemployed workers also tend to have higher rates of crime and violence."[8] Part of defining our values and morals is noting the ramifications of our choices as a society. If we are creating machines that are acutely intelligent, can they provide solutions to the problems of depression and poverty that they'll cause? Even if those solutions aren't immediately lucrative?

As a reminder, this portion of *Heartificial Intelligence* is dystopian by design. My goal is not to say that all AI is evil or money driven, or that robots are bad. My desire is to provide a greater context for these issues so we can analyze what values we seek to retain as technology encompasses our lives. Ethical challenges abound and provide us with real-world opportunities to hone and codify values we'd like to make a part of an AI ecosystem.

For instance, *Fast Company* recently reported how a series of technologies and innovations are helping fight Ebola in Monrovia, Africa, in an article called "Fighting Ebola with Robots and an App Called JEDI."[9] JEDI stands for Joint Electric health and Decision support Interface, referring to the effort to aggregate data about the epidemic via a standardized system for patients. The robot in the article is VGo's telepresence robot, which functions like a rolling iPad equipped with Skype for real-time interaction between doctors and patients. As the robot can be operated remotely and doesn't require human touch in the field, it has provided a great deal of support and a human face in the fight against Ebola. In this example, I'm more concerned that people don't die than I am about their personal data. If my kids were at risk for a disease of any kind, let alone Ebola, I'd happily welcome penis enlargement ads for the rest of my life to keep them safe.

However, if this same technology were employed within a self-driving car, I'd fight tooth and nail to protect people's personal data. It's only a matter of time before a service like Uber gets rid of human drivers in lieu of autonomous vehicles. To speed a rider's comfort and help him or her anthropomorphize a vehicle, I'm sure first iterations of these cars will feature a human face in a screen, providing a sense of camaraderie during travel. Cars will be equipped with every form of facial recognition and biometric sensor technology to maximize users' personalization experiences. Publicly available health and emotional databases will allow the car to prompt advertisements or other suggestions as part of the trip, or could even broadcast data to an employer if someone uses the vehicle as part of his or her commute. ("Hi Peter, your heart rate is off the charts and I'm worried your stress will affect our big pitch today. I've asked your car to take you back home.") Unless people control their personal data, all of these ethical decisions are out of our hands.

Engaging Ethics Around AI

> *You can't say it's not part of your plan that these things happened, because it's part of your* de facto *plan. It's the thing that's happening because you have no plan. . . . We own these tragedies. We might as well have intended for them to occur.*[10]
>
> —WILLIAM MCDONOUGH ON DESIGNING THE NEXT INDUSTRIAL REVOLUTION

I first learned about this quote from the book *Wired for War*,[11] by P. W. Singer, which focuses on the ramifications of militarized AI. He makes the point that while McDonough said these words in the context of ecological sustainability, they could just as easily apply to artificial intelligence. In his book, Singer proposes the idea of a "human impact assessment" required from any organization before production begins on an autonomous system or machine: "This will not only embed a formal reporting mechanism into the policy process of building and buying unmanned systems, but also force the tough legal, social, and ethical questions to be asked early on." I think this is a fantastic idea. While Singer is referring largely to militarized systems, I firmly believe this type of assessment should be required for any organization that utilizes evolutionary or machine-learning-oriented algorithms. In the same way organizations are held accountable for their potential effects on the environment, they'd now be responsible for their impact on their employees' well-being due to automation.

"Almost all of our laws are based on the underlying assumption that only humans can make decisions." In my interview with John Frank

Weaver,[12] we discussed how laws and policy are having to quickly catch up to AI technology that's already available in things like automated vehicles. "A lot of the new technology that's coming out is artificial intelligence and autonomous," notes Weaver. "It's different from robots and machines of the past because they can do analysis and judgment."

This analysis and autonomy distance the current AI environment from the type of distrust of technology that took place in the Luddite rebellion from 1811 to 1813 in England.[13] Skilled artisans fighting against the onslaught of industrialization and cheap labor is a far cry from concerns about how a self-driving car will be programmed in life-or-death situations. And as Patrick Lin notes in his article in *The Atlantic* "The Ethics of Autonomous Cars,"[14] the lack of any laws regarding nonhuman actors provides broad opportunities for companies like Google to push AI technology forward in an ethical and regulatory vacuum. The article quotes Stanford law fellow Bryant Walker Smith,[15] who noted Google's cars are "probably legal in the United States, but only because of a legal principle that 'everything is permitted unless prohibited.'"

Well, that's fun. Is it prohibited for me to build an autonomous U. S. Congress that actually gets work done? Or an algorithmically driven sundial the size of Denver powered by Nutella?

I joke, and yet I don't. These examples seem absurd, but some might feel going to court against an autonomous car manufacturer after a fatal car accident is equally far-fetched. But these are precisely the legal situations we're having to deal with today. And as Kate Darling,[16] of the MIT Media Lab, notes in her paper "Extending Legal Rights to Social Robots,"[17] "Long before society is faced with the larger questions predicted by science fiction, existing technology and foreseeable developments may warrant a deliberation of 'rights for robots' based on the societal implications of anthropomorphism." I'll talk more about robot rights in the next chapter, but the basic idea is that since corporations have been legally granted personhood,[18] these rights could extend to autonomous devices they manufacture, whether it's an Internet-enabled refrigerator, Furby, or a sexbot. For companion or "social" robots, designed to increase empathy or well-being in humans, there's a strong possibility we'll soon test our ethics regarding AI in the culture and the courtroom. We already feel emotionally attached to our cars—think how our auto-driven relationships will blossom when Siri-type assistants become part of their innards. Getting your car totaled or stolen will take on whole new repercussions akin to kidnapping, and we haven't figured out as a society how to deal with the ethical implications involved.

Fortunately, experts such as Darling are bringing ethical issues to the foreground in dealing with AI. In my case, I first seriously started thinking about this area after reading an article on the *Huffington Post*, "Google's New A.I. Ethics Board Might Save Humanity from Extinction."[19] It reported the news that Google had acquired the AI company DeepMind, whose cofounder Shane Legg had been quoted in 2011 as saying, "Eventually, I think human extinction will probably occur, and technology will likely play a part in this." He went on to say in a 2011 interview[20] that among all the tech that could destroy humanity, AI is "the number one risk for this century." To his credit, DeepMind agreed to their acquisition only after Google promised they'd create an AI safety and ethics review board. This aspect of the announcement garnered a lot of attention, from deep criticism and distrust of Google to a helpful elucidation of ethical issues around this type of board in an article from *Forbes*, "Inside Google's Mysterious Ethics Board."[21] And as I mentioned in the previous chapter, the Future of Life's AI petition appears to be the manifestation of this board.

T.Rob Wyatt is managing partner of IoPT Consulting,[22] whose mission is to "put people into the Internet of Things . . . so that the device owner is the first, and possibly only, owner of the device data." He's a thought leader in the personal data space whom I interviewed regarding his thoughts about AI ethics. In response to the idea of a robot, such as Jibo, that utilizes sensors to monitor people's activities within their home, he noted:

The XBox One Kinect was widely covered in the press for its ability to watch a room in incredible detail. It mapped the room in a 3-D mesh, accurately gauging distances and depths. It even has the ability to measure heartbeats of people in the room at a distance using color detection and infrared. When the Kinect was bundled into the XBox as an always-on, mandatory component, sales plummeted due to privacy concerns, eventually forcing Microsoft to unbundle the component.

Other than those who perform for pay, most people do not put a remotely controllable hi-def webcam and microphone in the bedroom. Certainly, people do not put them in their young daughter's bedroom. Yet, that is exactly the scenario that Jibo shows in their introductory video.[23]

Jibo has, at the time I'm writing this book, the "#1 most successful technology campaign on Indiegogo,"[24] according to the crowdfunding site. The campaign raised $2,287,110 for its $100,000 goal. Apparently people don't share the concerns that T.Rob and I have regarding the new iteration of surveillance-bots going into people's homes. I should also point out, however, as *Forbes* writer Ryan Calo notes in his article "Could Jibo Developer Cynthia Breazeal Be the Steve Wozniak of Robots?":

Never in a million years do I think Breazeal would permit her creations to be used in this way [manipulation by Jibo to inspire users to purchase specific products]. She has, as far as I can tell, devoted her life to putting humanity into human-machine interaction. Yet, is this true of everyone? If not, the issue is one bad actor away from a consumer protection problem.[25]

Dr. Breazeal[26] is director of the Personal Robots Group at MIT and a pioneer in the field of social robotics,[27] a vertical within AI research focusing on autonomous devices imbued with human characteristics or goals. Like Calo, I believe Breazeal and a majority of AI experts wish to use robots like Jibo to help humankind. Whether it's simply to eradicate tedium in our daily lives or teach us how to increase empathy so that we won't hurt our loved ones, I see enormous benefits with the implementation of these robots.

From an ethical standpoint, however, the ecosystem these cloud-based products are creating still operates within the confines of the established Internet economy. Breazeal and her team offer privacy and data protection for their device, but are also fast-tracking the negative actors who will piggyback on her technology or other systems like it.

T.Rob pointed me to a report produced by Northeastern University about the nature of personalization on the web. It measured the effects of testing price steering (manipulating products shown to a user) and customizing the prices of those products (price discrimination). Their experiments revealed that "based on our measurements from real people . . . several e-commerce sites implement price discrimination and steering."[28] Imagine how much easier it will be for companies to extend this price steering or discrimination after seeing into our homes. They'll be able to tell what we buy, how often we buy, and when we're at low emotional points where we'd be inspired to buy more.

The only way to maximize societal benefit from transformational devices like Jibo or Pepper is for individuals to control their own data. Otherwise, the Internet of Things personified by these robots or devices in our homes will be utilized for continued manipulation, accelerated by their widespread use.

Positive Psychology and Altruistic Algorithms

In the second part of this book I'll be discussing positive psychology, the empirically based "science of happiness" that focuses on how you can take actions such as expressing gratitude to increase your well-being. For most of us, studying happiness in our lives is left largely to chance. We wait until we feel a certain mood, typically depression, to ask ourselves questions like, "Is this job making me happy?" or "Is this relationship keeping me from being happy?" These are valid and common questions, many of which focus on something called hedonic happiness (from the same root word as *hedonist*), mentioned in the Introduction, which is the ephemeral mood that comes in response to emotional stimuli. Positive psychology shows that the pursuit of this type of happiness leads to something called the hedonic treadmill—an exhausting up-and-down array of intense emotions that can actually debilitate well-being rather than increase it. In opposition to this treadmill is eudaemonic happiness, often referred to as intrinsic well-being or flourishing. This is something you can increase by repeatable and measurable actions that you can test at your own pace. You can practice mindfulness to lower stress, or be altruistic to increase your self-esteem.

I believe the study of ethics around AI should be utilized to try and emulate the beneficial results positive psychology is having on well-being. Rather than pay lip service to ethics surrounding AI, we need to define and test the specific values we wish to incorporate into the machines taking over our lives. While we need not take an "us versus them" mentality, creating a sense of parity with our ethics just so machines won't annihilate us, it's imperative to specify how human empathy functions to a point where it can be encapsulated into code. Thankfully people in the field of social robotics as well as multiple AI experts are working to implement these types of characteristics in their work. But until the general public understands these issues, here are three top concerns elucidated in this chapter:

- **Robots don't have inherent morals.** At least, not yet. It's imperative to remember that programmers and systems need to implement ethical standards from the operating system level on up. Otherwise a properly operating algorithm simply seeking to fulfill its goals may pursue a course of harmful action.
- **Organizations need AI accountability.** P. W. Singer's idea of a "human impact assessment" provides an excellent model for society to pursue regarding organizations creating or utilizing AI. In the same way companies are responsible for the environment, this type of assessment would enable organizations to address issues surrounding automation, employee issues, and existential risk before these situations take place.

- **There are few to no laws for autonomous intelligence.** You have the right to remain silicon. As John Frank Weaver notes, all existing laws were written with the assumption that humans are the only creatures able to make decisions of their own volition. AI changes that dynamic, even with the weak AI being used today in autonomous vehicles. This need for new laws provides an excellent opportunity to define and codify the ethics that we want to drive our society, as well as our cars.

six | Bullying Beliefs

Winter 2022

Silence.

Please, God. Don't let them hurt her. Please tell me what to do.

Outside our apartment door the clomp of heavy footsteps echoed in the hall. My wife, Barbara, gripped Melanie to her chest, my ten-year-old daughter stifling a sob in her arms. Richard sat next to them on our threadbare futon, his hand on his sister's shoulder.

We waited.

A young woman spoke outside our door, telling her friend to check her mail, that she'd forgotten to get it last night. A deep male voice said to throw her keys so he could look. The keys jangled as she threw them and they fell to the floor. The woman laughed, her heels clicking as she made her way outside. A mailbox squeaked open and shut, the guy saying something indiscernible as the couple closed the front door with a *click*. Outside the building a cab sped past, horn blaring.

Silence again.

I kept looking through our door's spy hole to the street outside our apartment. In front of the building people were hurrying to work. A few kids were tentatively sitting on icy swings in the municipal playground between Forty-fourth and Forty-fifth Streets close to Ninth Avenue. I remembered pushing Mel there a few weeks back during a brief autumn warm spell. She still loved when I pushed her, still laughed when I made my baseball cap fly off my head as if she'd kicked me when I stood in front of her swing. I could picture her hair framed from behind by the golden-hour light that came at dusk. I'd smiled, said, "Hey, want to know something?" She'd rolled her eyes and responded, "That you love me?" I laughed, and she added, "Dad, you're such a *sap*."

But I did love her. More than my life. And now they were coming. I couldn't pay them and I didn't know what to do.

We'd left our house in suburban New Jersey about two years before. The absurdly high taxes on top of the mortgage were too much for us. We found an apartment complex in our town and cut our expenses by a third, but we weren't able to save a whole lot of money in the process with closing costs on the house. Moving sucked. We loved our neighbors, adored our house. I had always dreamed of a real fireplace, and we had one. The kids each had their own room; my wife and I had offices. It was perfect. Then we lost it.

Initially, automation had seemed like a natural evolution for society. Everyone I knew considered it part of the modern Industrial Age, in which Big Data was the new norm. A bunch of people might lose their jobs but it was just like every other generation had faced before. My friends applauded my being a "human writer" to sustain my job, laughed at the idea at first. Then many of them asked me privately how I'd pitched the idea to my boss, to see if they could make something along those lines work at their jobs. One friend was a well-paid medical technician with a number of degrees. Another friend asked for his wife, who'd been a highly compensated administrative assistant for years. They both lost their jobs in the same year. I wrote a few stories about this rapid increase in job losses until one of my editor-bots calculated the lack of positive sentiment in the tweets and posts that followed them. So those stopped.

We weren't the only family on the block who had to move. I took no comfort in that fact, although I felt a special bond with the other dads I saw packing boxes. Regardless of any issues of technological innovation or unfair circumstances, as husbands and fathers we felt we were failures. Perhaps from a cultural standpoint my son wouldn't feel this way when he got older, but as a forty-six-year-old who couldn't afford our dream house any longer I was emasculated. Shame filled my body like poison. It rose with me in the morning, kept me from sleeping at night, and accompanied me throughout the day. If you've never lived with debt, or job loss, or the constant terror of what those could bring, good for you.

But you're the exception.

It was only a few months after we moved into our thousand-square-foot apartment that I lost my job. Enough machines had taken over human roles that corporations felt comfortable transferring their legal personhood status onto the objects that housed their intellectual property. I'd seen this coming and treated Gigglepussy like a real person to avoid any legal issues. Yes, he was a real jackass of a "person," but ironically this helped me anthropomorphize him fairly quickly.

It was only when I stayed at the office late and the lights were turned off that I remembered he was a computer. I often got this weird feeling I wasn't supposed to be there, that GP was going to put the moves on our photocopier and I was cramping his style. Seriously, the weird feeling really happened all the time. Not the photocopier thing but the sense that GP was not the same type of creature as something like my pencil sharpener. Same thing with the 3-D printer we'd gotten after the MakerBot exploded in popularity. It was connected to the Internet and could even print human organs.[1] I'm not kidding—DNA from a donor was implanted in a plastic polymer reminiscent of foam rubber and you could bang out a liver. We didn't use ours to print any organs, but it was morosely reassuring that we could.

The Maker community that had formed in the 3-D printing community was amazing. So much creativity, so much freedom to print small runs of new products or devices that inspired entrepreneurship. That part of the story was inspirational. What was freaky was thinking about GP creating his own body to print out at some point, if he even wanted to. I read that a team in Oslo, Norway,[2] had created self-learning, self-repairing robots that could be 3-D printed for emergencies. So in the depths of a mine during a crisis, a robot could be printed in a zone too dangerous for humans. I saw no reason why cops wouldn't place machines like this in Times Square during New Year's Eve in case they had a crisis.

Print up a few dozen security bots if the tourists get too rowdy so Carson Daly can get back to work glorifying banality.

In the case of my job, print up a few new writers when you need to create content that will algorithmically rock the views. Management didn't actually need to create a humanoid body for me when I got fired, of course. They just dedicated a new algorithm to issues relating to humanity and began testing. By the time I left, algorithms had gotten so advanced that brands knew with 100 percent accuracy what would make people click their ads. Manipulation was so nuanced that brands actually had to keep from addicting people to their products to the point consumers went bankrupt. Not that they cared about the emotional aspect of people's losses, but if they went bankrupt they couldn't keep buying things. Analytics backed this claim.

What was more of a challenge was how brands divvied up and targeted each consumer, since the notion of genuine preference was now a thing of the past. Consumers were so segregated by demographics, so monitored by sensors, that any company could insert their product in real time into people's consciousness and guarantee a sale. Food products were the easiest. A dip in a person's blood sugar monitored by a wearable device signaled a dozen different companies, each one providing GPS analysis of the closest location to purchase their products. The six companies closest to the consumer participated in an instant lottery managed by the Digital Advertising Board. Designed to prevent monopolies (but often influenced by key lobbying efforts), the lottery chose two companies to directly pitch the consumer. The brand with the fastest algorithm won, much in the same way financial trading houses with faster AI[3] or closer proximity to Wall Street in years past secured the best trades.

The companies that pitched their ads to consumers rarely used print messaging any longer. Augmented reality contact lenses had been measuring people's retinas for years via their patented pay-per-gaze technology[4] as a proxy for emotion. It didn't matter if consumers felt they didn't like a certain product—the physical response of their bodies gave them away or provided clues about what would make them change their preference. Similarly, the earbuds most people now implanted directly in their cochleas[5] could route directly into the nervous system. Sounds that elicited specific emotional responses, highly tailored to each individual, would be sent at exactly the moment they considered taking a food break. The noise of crinkling paper in someone's office would be audibly modified just enough to be reminiscent of popcorn popping, as fresh and inviting as at the movies.

Some people had begun experimenting with brain hacking, using electric stimuli to target areas of their minds and experience specific physical sensations without the actual input of touch. For these fans of consumer neurotechnology,[6] advertisers could utilize local Bluetooth networks to generate audio electromagnetic blasts hyper-targeted to an individual's brain. The consumer being targeted would begin to crave a Coke, Lay's potato chips, or Reese's Peanut Butter Cups as if the process had happened naturally. Companies knew their triggers would produce a release of endorphins and hormones that took people's desires from mental to physical, from brain wave to brand crave. And then the individual would go and buy whatever it was she'd been manipulated to buy. The process wasn't a secret. It was the natural evolution of machine-learning algorithms, AI, and the advertising-driven consumer economy. Society had simply internalized the process that had begun online.

I received advertising messages all the time in response to my increased depression since I'd lost my job. Symptoms were easy to track via wearable devices and facial recognition tech. Simple algorithms could predict when I'd be most vulnerable to temptation. Targeted ads often entered my consciousness while I watched TV late at night, when I was doing my best to ignore our overwhelming debt. I put on fifty pounds in a matter of months via the massive onslaught of sugar and chemicals forced in my direction by brands outfitted with the real-time data of my low self-esteem.

We kept getting credit card offers, even though our rating had slipped. Apparently the cost of paper, stamps, and effort to tempt people suffering financial woes was more lucrative than offering them actual assistance. We maxed them all out in our first apartment in Jersey, one of the main reasons we moved back to Manhattan into our current five-hundred-square-foot abode, which we sublet from a friend who let us slip on rent.

Barbara and I couldn't find work. While automation experts said creative skills would be needed long after most jobs went to robots, my experience had proven the opposite to be true. Barbara had been a well-paid lawyer for years, but was also easily replaced by software programs that utilized brute-force algorithms to do research, filing, and billing at rates ten times faster than any human.

Our savings evaporated quickly and we could barely make rent on our new apartment. We'd stopped paying credit card bills months ago. Now the only people who called us were debt collectors. Every ring, every buzz on my phone with "unidentified caller" pushed layers of icy terror into my stomach. This was a process I didn't understand. The credit card companies knew how brands manipulated us to buy things. They saw the millions of people just like me who'd used food or pharmaceuticals or gambling to try and stave off the depression that had settled over our automated nation. Didn't they understand I couldn't ever pay them back? Hadn't their algorithms pinpointed the exact day and time when I would be completely unable to pay their minimum balances? Of course they had. But their bots kept calling me 24/7. They never got tired. They never needed a break. They never felt remorse or guilt about my plight.

They automated shame.

So today, we waited. Because the only thing of real value left in our house was about to be removed. It was the only asset worth the trouble of coming in person to our home, since creditors avoided sending robots into our neighborhood because they always got destroyed.

They were coming for Melanie's chip.

The general public had only just begun to start implanting software into their brains. While early adopters and transhumanists had been living with chips like Melanie's for years, they were the exceptions. But now cultural zeitgeist predicted "chipping" would be all the rage. Wearable technology had gotten people used to Internet-connected software becoming an extension of their bodies. People spent so much time on screens it seemed natural to bring the tech within, hacking optic nerves and neurons to experience the world through silicon.

Melanie's experience provided the rare case study of how a girl's prepubescent body interacted with the chip's synapse and electromagnetic brain relays. Her physiological data had been intricately tracked for years, mapped against adult women's to measure responses to everything from food to boys to trauma. Melanie's chip and the data it stored were a wellspring of intellectual property for its manufacturer. And now that we couldn't afford her latest firmware upgrade, they wanted it back. For the latest iteration of the chip, users never had to remove the unit once it was installed. All updates were made remotely via Wi-Fi and the web. But Melanie's chip had been a beta version of the technology. She'd already had two upgrades since she'd gotten the original. Both had required surgery and had taken years off my life.

But they'd saved hers.

We had no way of knowing if the childhood Parkinson's would return without the chip. Dr. Schwarma's best guess was that Melanie's brain functionality would continue as normal once it had time to repair itself after surgery. But I knew Dr. Schwarma—knew how much she adored Melanie. I'd seen the tightness in her face when she'd delivered the news to us. She wasn't lying with her prognosis, but she also wasn't elaborating on potential negative outcomes. To be fair, there was no evidence of what would happen if you removed a chip from a healthy patient like Melanie. Nobody would ever do such a thing.

Unless you were me, and you didn't have a job. Unless you were a man defeated. By machines. By circumstances. By the way the world operated.

I looked at my family. They looked up from the couch. Melanie pulled away from Barbara's chest, wiping tears with the back of her hand.

"Mel," I said.

"Yeah, Dad?"

"Want to know something?"

She smiled, sniffed. "I love you too, Dad."

The buzzer for our apartment blasted near my ear. Three quick, impatient beeps. I looked through the spy hole, saw two men in matching uniforms standing outside on the street, one of them holding a medical bag. After a moment, they buzzed again, this time one of their voices coming distorted over the intercom:

"Mr. Havens. We're here for your daughter. Open the door."

> *If we don't treat cyberconscious mindclones like the living counterparts they will be, they will become very, very angry.*[7]
>
> —MARTINE ROTHBLATT, PH.D., VIRTUALLY HUMAN: THE PROMISE AND THE PERIL OF DIGITAL IMMORTALITY

The Gotham Writers' Workshop explains that the difference between science fiction and fantasy in literature is that "science fiction explores what is possible (even if it's improbable), while fantasy explores the impossible."[8] They quote Ray Bradbury, who described science fiction as "sociological studies of the future, things that the writer believes are going to happen."

Bradbury's description aptly explains where I live most of my mental life. I'm insatiably curious and am fascinated by science and technology. I like fantasy—I'm a huge fan of *The Lord of the Rings* and other works from that genre—but there's something especially compelling about a story that features a twist on our lives that could actually take place in the near future. This is how I've worked to create the opening narratives to the chapters in this book, to demonstrate that the majority of tech and trends I'm describing are *already here.* They're not science *fiction* as much as science *inflection*, in that they exist and are at a point of cultural discovery. Their adoption has started but is not yet widespread.

I'm not a fan of scare tactics or link baiting, and as you'll see in the second half of this book, I believe we have multiple opportunities to free ourselves from the artifice of manipulative autonomous intelligence. A lot of this freedom will come in the form of personal data control, but it also relies on us taking the time to identify the values we want to codify for the devices permeating our lives. Make no mistake, this is tough stuff to figure out. Questions like "What does it mean to be human?" have become a lot more complex since we started looking for our answers via the Internet instead of within. Where we stop and machines start isn't clear any longer. By the time we're looking at copies of ourselves in robotic form the mirror is going to be cloudier still.

The Bully in the Bits

Martine Rothblatt is the highest-paid female CEO in the United States,[9] a renowned lawyer, futurist, and businessperson featured in 2014 on the cover of *New York* magazine in the article "The Trans-Everything CEO."[10] Born male, she underwent gender-confirming surgery in 1994 and remains married to her life partner, Bina Aspen. The couple wed more than thirty years ago, while Martine was still a man. Rothblatt also created a robotic version of her wife called Bina48, which features a highly realistic face along with multiple technologies such as facial recognition, voice recognition, and AI.

You can learn more about Bina48 at the LifeNaut[11] (like astronaut) website, and also create a mind file like the one I described in Chapter Three for my fictional robot son. The LifeNaut project is part of the Terasem Movement Foundation, whose managing director, Bruce Duncan, M.Ed., spoke at the Ideacity conference in 2013 while providing a demo of Bina48. Duncan pointed out that the concept of mind files will be as transformative to our species as the time humans first created language. He notes, "We started sharing stories as a species by creating cave paintings in France. Now the Internet is connecting all our neocortexes." LifeNaut is facilitating this process by helping people to upload an expression of their consciousness via digital files such as photos, videos, and posts. In this way, the memes of our lives will eventually allow our digital doppelgängers to be in multiple places at once. As Duncan states:

> *If you're the steward of your own mind file, artificial intelligence can reanimate and bring us to other places, like how we use a telephone—*

we're in two places at once. In the near future, say, ten to fifteen years, we'll think nothing of the fact that something is backing US up. Our attitudes, our beliefs, the essence of us.[12]

While I'm still wrapping my mind around the concept of multiple copies of myself traversing through time and space, the idea makes a lot of sense inasmuch as that's what's happening to our personal data right now. Copies of my identity are currently quite granular and controlled by dozens of disparate organizations, but they exist. I would much prefer to tie them together in a safe repository like the one offered by LifeNaut, where I can add and curate my own identity versus leave it for other actors focused primarily on profit.

While I applaud LifeNaut's philanthropy (as of the writing of this book, it's currently free to create and store a mind file at www.lifenaut.com), I have some issues regarding Rothblatt's vision as she outlines it in her book. My first concern comes in her quote about how we should treat cyberconscious mindclones. In the context of *Virtually Human*, this is a theme Rothblatt comes back to a lot. She provides the compelling analogy of negative treatment of mind files being similar to blacks dealing with racism and the gay community dealing with homophobia. But if these mindclones are utilizing AI, won't they be able to ignore our stupid human pettiness? I realize we'll imbue them with emotions, but won't they realize as first-generation mind files that we humans will require some time to adjust to the *biggest transformation of the human species since we evolved from protozoa*?

I'm also not thrilled Rothblatt felt the need to repeat the word *very* when describing how potentially angry these mindclones will be if we don't "treat them as the lifelike counterparts they will be." It's a threat, plain and simple. How are we supposed to acclimate to the idea of these new forms of life at virtual gunpoint? And precisely how would the anger of these mindclones manifest toward humanity?

Just as troubling is Rothblatt's focus on the monetary aspect of mindclones. While this is an understandable focus due to her success as a businessperson, it negatively colors my understanding of these future entities who will supposedly be our equals. Here's a quote from *Virtually Human*:

> *Digital cloning of our own minds is being developed in the free market, and on the fast track . . . vast wealth awaits the programming teams that create personal digital assistants with the conscientiousness and obsequiousness of a utopian worker. As uncomfortable as it makes some—a discomfort we have to deal with—the mass marketing of a relatively simple, accessible, and affordable means for Grandma, through her mindclone, to stick around for graduations that will happen for several decades from now represents the* real *money [her italics].*[13]

Creating an "obsequious utopian worker" hardly feels like equality to me. Realizing there will be various levels of mindclones and the robots they inhabit, if they're imbued with any form of consciousness, won't they be angry at being made to do our bidding? And won't mindclones with higher levels of consciousness feel for the less endowed members of their race, as the fictional me pointed out to my robot son in a previous chapter?

It's also difficult for my anticorporate skepticism not to kick in at Rothblatt's focus on the "*real* money" that will come via the ubiquity of mindclones. While I appreciate she wants to make this technology affordable for everyone, if we are to treat mindclones as legally represented, holistic entities, where does their sale not equate to the machine equivalent of human trafficking? Her noteworthy stance on avoiding the evils of "fleshism," or racism directed toward sentient mindclones, loses credibility when she notes the ease with which they'll be sold and utilized for our every whim.

Overall, the aspect of Rothblatt's vision I find most challenging is her sense of inevitability that mindclones imbued with AI will soon overtake humanity. While I actually agree that the ramifications of autonomous intelligence are inevitable and are already upon us, it's her technological determinism (the theory that a society's technology drives the development of its social structure and cultural values) I find most concerning. It mirrors a pervasive attitude within Silicon Valley that you should produce a certain technology because you *can*, not that you *should*. Any ethics or values involved in innovation only come into question *after* a technology has been introduced to the marketplace. Therefore any discussion of values revolves around genie-back-in-the-bottle imperatives versus creating technologies that everyone agrees will bring value to humanity.

Along these lines, Jürgen Schmidhuber[14] is a computer scientist known for his humor, artwork, and expertise in artificial intelligence. As part of a recent speech at TEDxLausanne, he provides a picture of technological determinism similar to Rothblatt's, describing robot advancement beyond human capabilities as inevitable. He calls the time when forty thousand years of "human-dominated history" will end Omega, a term equivalent to the Singularity, which he predicts will occur around the year 2040. In his talk he observes that his young children will spend a majority of their lives in a world where the emerging robot civilization will be smarter than human beings. Near the end of his presentation he advises the audience not to think with an "us versus them" mentality regarding robots, but to "think of yourself and of humanity in general as a small stepping stone, not the last one, on the path of the universe towards more and more unfathomable complexity. Be content with that little role in the grand scheme of things."[15]

It's difficult to comprehend the depths of Schmidhuber's condescension with this statement. Fully believing he is building technology that will in one sense eradicate humanity, he counsels nervous onlookers to embrace this decimation. Later in his talk he also makes the very salient point that almost no politicians are aware of the rapid increase in AI technology and its cultural consequences. So the inevitability of our demise is assured, but at least our tiny brains provide some fodder for the new order ruling our dim-witted progeny. Huzzah! Be content!

This is not a healthy attitude.

For the programmers, ethicists, and social scientists who understand the ramifications of impending machine autonomy, it's imperative not to usher in our potential human evolution in ignorance. You can't actively build a framework of systems that can automate jobs, emotions, and relationships while ignoring the humans you're replacing.

Unless, of course, you're a bully.

This is precisely what David Gelernter,[16] professor of computer science at Yale University, in his article for *Commentary* magazine, "The Closing of the Scientific Mind," calls the scientific determinism exhibited by Schmidhuber in his talk:

> *Many scientists are proud of having booted man off his throne at the center of the universe and reduced him to just one more creature—an especially annoying one—in the great intergalactic zoo. That is their right. But when scientists use this locker-room braggadocio to belittle the human viewpoint, to belittle human life and values and virtues and civilization and moral, spiritual, and religious discoveries, which is all we human beings possess or ever will, they have outrun their own empiricism. They are abusing their cultural standing. Science has become an international bully.*[17]

By proselytizing the inevitability of the Singularity, we lose the once-in-a-species opportunity to define what makes us human less the pressure of our imminent demise. Sentient AI hasn't arrived yet. As Gelernter points out, it's a form of genocide to extinguish the unique beauty of the human mind and spirit to make room for highly probable fiction. Relinquishment, the attitude that all AI research should stop as a means of saving humanity, is unrealistic at this point of advanced autonomous research and implementation. But the difficulty of creating ethical standards doesn't mean we shouldn't try.

We're worth the effort.

Hungering for the Human

> *One of the most fascinating elements of the rise of machine intelligence is what it highlights about humanity. We are creating*

these machines, and as such, we are making them in our image. Some say that we choose robot relationships because we have more control, but I think inherently we are drawn to the messy, to the complications of love and emotion. That's what makes us human. The reason we have these companion bots—elder care bots, nanny bots, romance bots—is that we've become so busy. We're so caught up in this culture of being busy that we don't have time for each other. So it has to be asked, are the robots becoming more like us, or are we becoming more like robots?

—RAMONA PRINGLE, CANADIAN DIGITAL MEDIA NETWORK JOURNALIST AND CREATOR OF AVATAR SECRETS

My friend Ramona is a thought leader in the world of gaming, transhumanism, and digital culture. Her iPad experience, Avatar Secrets, combines live-action video footage, animated sequences, and interviews with multiple experts in the digital arena at large. After Ramona experienced a particularly difficult time in her own life, with the loss of a long-term relationship and a loved one's illness, she retreated into the world of massive online games to find comfort. Her experiences led to an exploration of what individuals and society may feel they're missing that leads them to delve into virtual worlds.

It should be noted that gaming or artificial worlds can provide solace that many people don't experience in their day-to-day lives. As game designer Jane McGonigal points out in her TED speech[18] and her book, *Reality Is Broken*,[19] immersive games provide people heroic pursuits with live players in the form of avatars. The sense of increased self-esteem and benefits of team achievement are genuine in these environments. Likewise, a recent *New York Times* article, "To Siri, with Love," documents the poignant story of a thirteen-year-old autistic boy's connection with Siri, the personal assistant available on iPhones. While Siri wasn't developed to directly deal with autism, it brings comfort and happiness to the boy.

These examples demonstrate the positive opportunities technology provides in the realms of virtual and AI technology. And in the case of gamers and the parents of autistic children, it's their personal choice to deal with technology in the ways they choose. It is not my role or my desire to judge their choices and I'd be a hypocrite if I did, due to the dependence on devices in my own life.

But the availability of these technologies doesn't justify our dismissal of each other. Siri seems like a fantastic tool to aid the study of autism. But as a parent in the same situation, I would hope my child could leverage that tool to increase the value of her human relationships. Gaming is awesome, but it shouldn't keep us from trying to create a real-world environment providing purpose and community.

We're allowed to celebrate our humanness apart from the wonderful tools we've built. We're allowed to recognize the differences between creator and created. Otherwise, we accept the artifice that the intelligence we build defines our worth, denying the beauty of our humanity as we exist *right now.*

Here are the primary ideas from this chapter:

- **The Singularity already exists as an objective.** The ideas, philosophies, and economic imperatives driving multiple areas of artificial intelligence exist today. While AI experts may believe sentient autonomous technology is decades away, it's an existential threat that needs to be dealt with now. To deny its importance is to accept its potential consequences.
- **Mind files and money.** Our digital doppelgängers already exist. We can control them with programs such as LifeNaut or let them be curated for the benefit of advertisers, data brokers, and preferential algorithms. There's no middle ground.
- **Separation of search and state.** Scientific determinism is a philosophical choice, akin to religious faith. The belief that autonomous technology will evolve humanity to a lower or lesser state demands legislation, regardless of the benefits that the technology may or may not bring.

Toward Purpose and Progress

There's a concept in ethics known as moral absolutism,[20] which I've found very helpful over the years. The basic idea is that certain actions are simply right or wrong, or "good or evil." I first learned about the concept after reading C. S. Lewis, famed British atheist turned theologian. To define the concept, he provides the example of a commuter on a bus and how she'd feel based on two different scenarios. In the first, the commuter is walking to get a seat and someone closer to the seat takes it first. In the second scenario, as the commuter is about to sit, a person knocks her out of the way to take the seat. Beyond the potential for physical pain, why is the commuter likely to be angry at being denied a seat in the second example? Lewis posits there's a natural, innate understanding in this individual of what is right and wrong that raises her natural ire. It is also a humanistic trait versus an animalistic instinct, since in this example not getting a seat won't deny the commuter her life.

The opposite of moral absolutism is moral objectivism,[21] in which context or consequences also play a role in the ethical actions of an individual. So, for instance, if the person bumping the commuter were handicapped or had suffered a stroke, we'd feel he was justified in his actions. Or if someone steals to feed his family, this would be seen as acceptable. These circumstances make sense to me. However, I'm an absolutist when it comes to a number of issues, such as violence toward children. I feel violence toward children is wrong. If you told me of a religion that required child sacrifice or encouraged parents to beat their kids as a form of discipline, I would define those practices as morally abhorrent and ethically unacceptable. I'd work with others to create legislation to keep kids from being hurt. If I knew of a culture that encouraged sexual assault of women, that's also something I would judge as wrong.

While it's laudable to aspire to political correctness in modern society, it's also critical to draw boundaries around what we as humans feel is universally unjust or unwise. We've come to a time when technology is replacing us. But unlike the Luddites, the threat is not just about our jobs, but about our lives. While some may take solace in being a stepping-stone for the natural evolution of humanity, others recognize this as the rhetoric of redundancy.

But here's the good news: This is the end of the dystopian portion of *Heartificial Intelligence.* As I'm going to point out, a great many minds in the AI community and the world at large understand the threat of ignorance and fear autonomous technology has already created. Experts in the world of academia, business, and government are breaking down silos to define the values humanity wishes to embrace to enjoy the benefits of technology without the threat of termination. It's a sacred opportunity to pursue long-term goals versus short-term gains. It's time to embrace genuine progress without relying on machines to make the decisions that will drive our future.

SECTION TWO

GENUINE PROGRESS

seven | A Data in the Life

Winter 2023

I used to begin every day on Facebook. I'd come downstairs to make coffee for my wife, rushing to check my feed before she sat down so I could see what important messages or news might have accumulated while I slept. Oftentimes I got caught up in some silly video or angry screed that kept me from spending a few extra minutes talking to my wife before we started our day.

The habit carried over into mealtimes with my kids, especially if we went out to eat. During any lulls in the conversation, I'd feel justified in surreptitiously nudging my phone to see if I'd gotten any new messages. If someone new had just followed me on Twitter, I'd click on his profile to scrutinize his tweets. If a new e-mail appeared from one of the List-servs I followed, I'd check out the latest post from one of my fellow geeks.

Until one day Richard and Melanie asked me to stop.

"Dad, you promised you wouldn't check your phone while we were out together as a family," eight-year-old Richard said while we were out to eat on New Year's Day. "That's the resolution you made to us," he added, indicating Melanie, who was head down, drawing on the table-cloth at our local pub with a fistful of crayons. "A resolution is the same thing as a promise, right, Dad?"

Well, that got me.

Not checking my mobile first thing in the morning became one of the hardest challenges of my adult life. But I was big on keeping promises to my kids. So after I saw Anderson Cooper reporting on a headset[1] that could measure stress by analyzing brain wave patterns, I bought one to try it out. Every day for two weeks I tracked my stress and anxiety as I checked Facebook and my social networks. The following two weeks, I had Barbara hide my phone until after we'd had coffee together. I did okay the first two days but by day three began searching for my phone while she was still asleep. I rifled through her desk drawers, rooted in the cereal cabinet. I searched so long I forgot to get the kids up for school and they were both late.

This is when I realized I had an addiction to technology. I'd known academically I had it for years—I wrote about it for my work, after all. But it's different when you confront it. It's like the first time you get back to the gym after not working out for a long time. It sucks to see the folds of your gut outlined in your T-shirt. It's condemnatory to stand on the scale, fudging with the squeaky gym weights to coax a half pound in your favor. But in my case, I'd put the weight on. I'd made choices to avoid being active and ignore my health.

And I'd made choices to prioritize technology over my family. That weighed heavier on me than my flab.

Technology had become such a part of all our lives in society, it was challenging to show people how abusive it had become regarding our well-being. Like alcohol, it was beneficial in moderation. But until a few years back, the surveillance economy condemned outright digital abstinence. While Google and other Silicon Valley companies encouraged meditation and mindfulness[2] as part of their company cultures, many felt this messaging was an attempt to cover up widespread data-profiling practices. Cynics like me even felt companies pushed the meditation to provide employees with tools to deal with their guilt at manipulating the masses.

But then something changed.

Inspired by the ideas of Chade-Meng Tan[3] (Google's "jolly good fellow" and the author of the *New York Times* best seller *Search Inside Yourself*[4]), a young engineer at Facebook named Rebecca began practicing meditation as part of her daily regimen before going to work. Initially skeptical of what she secretly felt might be "trumped-up hippie bullshit,"[5] she realized the Silicon Valley version of mindfulness was simply brain training. She spent an hour a day at the gym, so thirty minutes a day spent honing her mind made perfect sense. A friend told her about the website Lumosity, and she signed up right away, enjoying the concept of playing personalized games that would sharpen her cognition. Within a few months, she noticed a profound drop in her stress and felt she was gaining a deeper sense of emotional intelligence as well.

But it wasn't until Rebecca read a quote from Vietnamese Zen Buddhist monk Thich Nhat Hanh[6] that she made a decision that would change the course of humankind. Clicking on a friend's tweet that led her to the website Tiny Buddha,[7] she read Hanh's words: "The most precious gift we can offer anyone is our attention. When mindfulness embraces those we love, they will bloom like flowers."[8] This concept blew Rebecca's mind. Or technically, it blew her mindfulness. While she'd been attracted to meditation as a tool to help herself, now she wanted to share the benefits of her enlightenment with others. She also felt it was the worst type of hypocritical pandering to espouse meditation while hijacking the identities of millions of users to mine their data.

Rebecca requested a meeting with Mark Zuckerberg and shared her concerns about the dangers of capitalism, how consumption and social status were not sustainable and brought about human suffering. Zuck had appeared nervous and bored during their conversation until Rebecca also pointed out the business opportunity in allowing people to own or control their own data. She offered to build the prototype for a cloud-based storage service to compete with tech rivals such as Dropbox and Amazon. Users could be charged to put all the data they'd created on Facebook and all their other social networks. Even her rudimentary financial figures showed massive profits.

That caught Zuck's attention.

Facecloud wasn't an instant hit. It took two years after beta testing for users to believe Facebook was genuine in its desire to offer users control of their data. Fortunately, Rebecca had brought MIT Media Lab's Alex

"Sandy" Pentland[9] into the project after she'd read about his New Deal on Data,[10] a "rebalancing of the ownership of data in favor of the individual whose data is collected, [so] people would have the same rights they now have over their physical bodies and their money."[11] Pentland's concept was fairly simple: In the same way an individual has a dashboard for a wearable device such as a Fitbit, the New Deal concept widened the scope of the dashboard to include the Internet of Things. Now entities from any source requesting data from an individual had to let him or her know what information they wanted to mine, and how it would be used. Pentland's ideas had been tested in Trento, Italy, where hundreds of families abided by the New Deal on Data they'd created as a community. As Pentland pointed out in an interview with *Harvard Business Review*, "These people share a lot more than people who don't live under New Deal rules, because they trust the system and recognize the value in sharing. Being confident about your personal data makes for a better economy, not a worse one."[12]

These ideas helped Facecloud evolve into a highly personalized economic dashboard versus merely a safe repository for people's data. It took some getting used to at first. While some users didn't like the idea of providing Facebook with a "golden copy" of their data, or an encrypted version of their PII (personally identifiable information), Facecloud integrated Facebook Connect into their service so people could traverse the web safely. Users knew their information was protected, as the service only parceled out the minimum amount of preapproved information needed for any transaction.

This meant a person could opt out of being tracked on any website he or she visited throughout the day. Cookies or other tracking devices were recognized and deleted by Facecloud before they generated any information about a user. The same logic applied to the Internet of Things. When a person using Facecloud went to a house using a smart thermostat, the user could opt out of having his or her heat signature tracked. This led to new forms of cultural etiquette, in which Facecloud users would let their friends know where they blocked certain forms of tracking.

For people who didn't care about data tracking, nothing changed. The people who did care finally had a choice.

Eventually Google designed their products to integrate with Facecloud. They'd been losing revenue from users who wouldn't get in Google's self-driving cars, which featured deep surveillance such as facial and voice recognition technology. So Google simply offered a choice: Facecloud users or anyone else could pay to ride in their cars without surveillance, or people could "pay" via preapproved usage of their data. Some Google fanboys complained the architecture brought complexity where it wasn't needed, since in the current system people's data paid for services by advertising anyway. But people craved the transparency that Facecloud provided, not just for the philosophy of openness but because of the clarity around how their data was being used. For the first time since the creation of the Internet, individuals got to see how often their personal information was being utilized for their own insights and gain. Now they could benefit directly from these assets.

Initially, Facecloud got a lot of attention from privacy advocates and fans of the open-source movement. But it wasn't until Facebook incorporated technology from Oculus Rift for users to experience how their data was mined in real time that the service exploded. Initially offered as bulky headgear reminiscent of safety goggles for eye protection, Oculus had dropped in price and size a few years after its initial release. Now sold in a box a few inches larger than an iPhone headset, Oculus consisted of two contact lenses and a matching set of earbuds. Wearing the gear, anyone looking at you would think you were listening to music or speaking on the phone.

The product, however, was equipped with a combination of Bluetooth and Beacon technology that sent visual or auditory alerts anytime another system tried to access your data. Scores of YouTube videos had come online within days of the new Oculus-Facecloud system becoming available, with screencaps of people's experiences as they walked through the streets where they lived. The screencaps showed how in the upper-right-hand portion of an Oculus-Facecloud user's vision, a red flash accompanied by a small chiming sound signaled attempted data access from an outside source. The highest rated YouTube video in this category showed a user getting her data accessed more than seven thousand times walking two blocks from her flat in London to a grocery store. The flashes and the chimes made the short walk seem like a battleground. This level of data distraction was the proving ground for Facecloud. People finally could visualize the depth of tracking in their lives and craved control of the multiple data exchanges they faced every day.

Facecloud began offering a range of options for users depending on what information they were seeking or what services they desired in various situations. For instance, in most public places individuals wanted to use their GPS or mapping apps, and Facecloud indicated what data they'd need to provide nearby services to make that work. In many locations, free Wi-Fi came with malware—programs that inserted tracking devices into a person's mobile for criminal purposes. For Facecloud users, this meant they either needed to wait to use safe public Wi-Fi, switch to roaming capability, or use static maps on their phones. Some individuals—shockingly—even took to asking other humans directions in these scenarios.

After the more utilitarian aspects of the system were set up, a Facecloud user could control his or her social settings as well. This meant in public settings a lot of people walked around with a Twitter or LinkedIn symbol appearing to float above their head as seen through Facecloud's digital lenses. Or they'd broadcast an envelope icon instead, indicating they wanted to be contacted before releasing any information in public.

In my case, I let people access my social networks or send me e-mails and texts when I'm in public. However, if my Facecloud system recognizes someone is using facial recognition software to identify my image for a photo, it blurs my face to keep it from being recognized. The law doesn't prevent anyone from blocking their image from being tagged without consent. This helps people control where they're identified in the public arena, both virtual and physical.

I also project my well-being scores in public. Like the digital communities that formed around Fitbit and other exercise apps, a number of people are now part of groups centered around services that improve well-being. For instance, you can't walk ten feet wearing Facecloud-Oculus without seeing someone with an icon featuring a picture of a human brain. That means he's a fan of mindfulness. Other people broadcast icons of a smiling face with two small hands facing outward. This means people are practicing gratitude. When I'm feeling gregarious, I switch on my real-time emotional meter to portray my well-being as colors on a wheel, shifting in intensity throughout the day. It's essentially a mood ring on steroids and is also available as a colored aura that radiates around my head and shoulders.

It's fairly crunchy granola, I admit. But it's a pretty powerful way to live.

My favorite part of this well-being framework is not the fact that people can see the projections of my inner life. Rather, it's the fact that I get indications in real time when there are Facecloud users nearby whose well-being could be improved via the skills and talents I possess. It's kind of like wearing a name tag at a networking event, but instead of getting a new client I get a new friend. In my work studying altruism, I've learned that helping others increases a person's self-esteem. So now Facecloud is helping users increase their well-being while honing their interpersonal skills in the process. The platform provides a massive opportunity for people in a community to help one another without having to feel invasive, as users choose when to make their needs public. For individuals who try to game the system, users provide commentary or feedback, creating a ratings system based on accountability and trust.

Facecloud has created an ecosystem of apps so things like my well-being platform can be correlated with the economic aspects of a community. For instance, people's tax statements can be broken down to indicate areas where they could help their neighbors to reduce their bills. Many communities have also created volunteer databases so people can visit the elderly to boost their own well-being while helping the infirm. The increased benefits of these types of visits have led to a dip in health care and pharmaceutical costs for depression, along with tax breaks for citizens in communities utilizing the system. Similar frameworks have been set up around education, physical health, and environmental issues. By offering time and talents catered to specific community needs, citizens increase their well-being while providing specialized resources local governments don't need to pay for.

Facecloud is truly remarkable, as is the New Deal on Data, upon which it was built. And while I love all the new apps that are available for the system, some days I opt to have it block out *all* data so I can see the world as I did before the Internet. Before augmented reality. Before artificial intelligence.

My son, Richard, has a simple solution for this as well. "Take off the lenses and the earbuds, Dad. Enjoy real life for a change."

And now that my data is protected and in my control, I can.

> *So far, personal data has been fragmented into dozens or hundreds of verticals. Soon it will be possible to collect a kaleidoscope of those slices into a single database to assemble the world's first digital representation of your life in unprecedented breadth and depth. The value of that dataset will be irresistible to any vendor who has it. The only person who can manage that data responsibly is the person who the data represents.*
>
> —T.ROB WYATT, MANAGING PARTNER, IOPT CONSULTING

Privacy isn't dead. It's just been mismanaged.

Privacy is well defined outside of discussions around personal data. This is why most of us would get angry if a stranger walked into a bathroom we were using and opened a stall to begin a conversation. Or we might take issue with someone passing out flyers for an auto dealership at a friend's funeral. We'd call these things invasions of privacy.

In digital circumstances, many of us have allowed ourselves to think that privacy is dead, as espoused largely by the organizations who profit from the proliferation of other people's data. Recognize the source. From a legal standpoint, depending on where someone lives in the world, privacy is certainly not dead. It's legislated in many different forms, and people around the globe fight fiercely to protect it in all its manifestations.

I choose to discuss issues around personal data using the word *control.*

In this sense, we can work to create systems that allow individuals to make their own choices regarding their data, philosophies on the preference regarding privacy notwithstanding. Think about money, for instance. The institution of a bank allows people who use its services to have a safe place to store and exchange their money. If an individual or organization outside this trusted relationship (banker-customer) tries to access a person's money, this is what is deemed an invasion or a crime.

Banking provides a solid analogy regarding an individual's personal data. Each piece of information you share about yourself is an asset. Sometimes it might be something with monetary value, like your credit card information. Other times its value comes in the form of an insight about your life, like your current location.

But here's where the banking analogy gets more complex. While people typically place their money in just one bank, data shared in today's Internet economy is parceled out to hundreds of organizations in the course of just one day. Some of these companies have direct relationships with their users, like Facebook or an online retailer. But the majority of entities grabbing shards of your identity have no relationship with you. You've made no direct contact with them to co-identify a value exchange. This means they're benefiting from your actions and identity without you receiving any compensation in return. In short, they're stealing your assets. Like a thief at a carnival, they nick your wallet while you're looking in a different direction.

This is the artifice regarding the intelligence these entities take from you—their insights exist outside the context of a consensual relationship. Picture a guy asking a woman for her number, and he thinks she's really into him. She smiles and hands him a napkin with a fake number, which he calls the next day. He misinterpreted her actions, basing his decisions on false assumptions. Multiply this example a thousandfold and this is the current state of the Internet economy. While the algorithms identifying people's behaviors and preferences continue to increase in complexity and nuance, without the transparency of mutual consent they fall short. Brute-force analysis pales in comparison to the value created within a trusted relationship.

Two Models of Management

In the current Internet (and Internet of Things) ecosystem, our personal data is mined and analyzed within hundreds of company-controlled customer relationship management (CRM) systems. These systems track our behavior to look for insights that will inspire purchase. As you can see from the diagram below,[13] a difficulty with the concept of CRM is that organizations tracking individuals all utilize different systems to gain their intelligence about you. They track within the context of what they sell, typically without a direct relationship to a user.

Vendor relationship management (VRM), however, transforms this dynamic. Created by Doc Searls, author of *The Cluetrain Manifesto*[14] and fellow at Harvard University's Berkman Center for Internet & Society, Project VRM seeks to "encourage development of tools by which individuals can take control of their relationships with organizations—especially in commercial marketplaces."[15] The project features a robust wiki with information posted since 2008 from around the globe.

Take a look at the diagram featuring VRM. As you can see in the picture, the individual controls how her data is accessed by any other individual or organization. While the diagram shows a picture of a computer, most people discuss this safe exchange portal as a personal cloud.

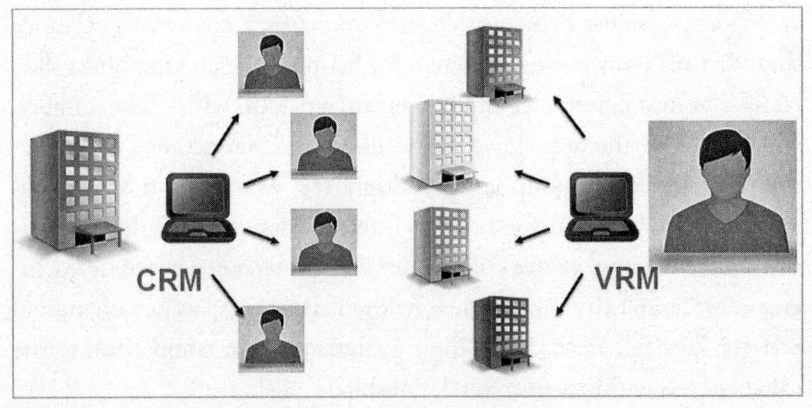

Caroline Condamin

It's also called a data bank, mirroring the analogy I provided earlier. Whatever its name, it acts like a Dropbox or other service that stores your data safely, as if it were on your own secure server at home.

If you worked in an office in pre-Internet days, as I did, you'll remember typing up a Word document and storing it on a floppy disk or USB stick to give to a colleague. The only copies of that document were on your computer and with people who were supposed to have it. You might worry about misplacing your disk or damaging it and losing your data. But unless you worked in the espionage business it was unlikely someone would steal it without your knowledge or permission.

Now picture yourself in today's world. By the time you've picked up any standard mobile phone, you've already revealed your physical location to any number of apps or services to whom you've given access. Check Facebook and any click is tracked and fed to algorithms serving you sidebar ads. Visit a new website? Hundreds of cookies instantly track your behavior, many with possible malware to insert on your computer. Walk outside and your fitness tracker information can be sent to its manufacturer. Shop in any retail store and Beacon proximity technology can intimately recognize your behaviors. Use unencrypted Wi-Fi at a Starbucks and hackers sitting nearby can break into your computer and steal your data.

Any of these events happen in all our lives on a daily basis, dozens of times over. Each company tracking us produces a different picture of who we are for analytical purposes. These myriad identities populate databases around the world, none of which we can access. Tiny fragments of who we are. Constellations of insights about our lives, owned by others. Never seen by us.

While I do believe data brokers and other organizations proactively keep this system opaque to enhance their profits, a majority of advertisers and companies track our data as they do now because it's the only system available to them. They track us because it's the only way they can gain certain information about our lives. This doesn't mean they shouldn't work to implement a better paradigm, however. But as Doc Searls noted to me in a past interview for Mashable:

We've digitalized the Industrial Age. Companies feel they have to get to scale so they have to treat you the same way. But people want to be treated as individuals. This should be a base condition for any functional marketplace.[16]

Do you have a local store that knows who you are? I have a couple. One is a breakfast place, where the staff knows I like soy milk in my coffee. It's a weird yet unique thing about me they've made the effort to remember. I'm both touched and impressed they've done this, and I reflect my appreciation in the tips I leave them. This is the type of relationship people should have with any of the organizations seeking to utilize their data.

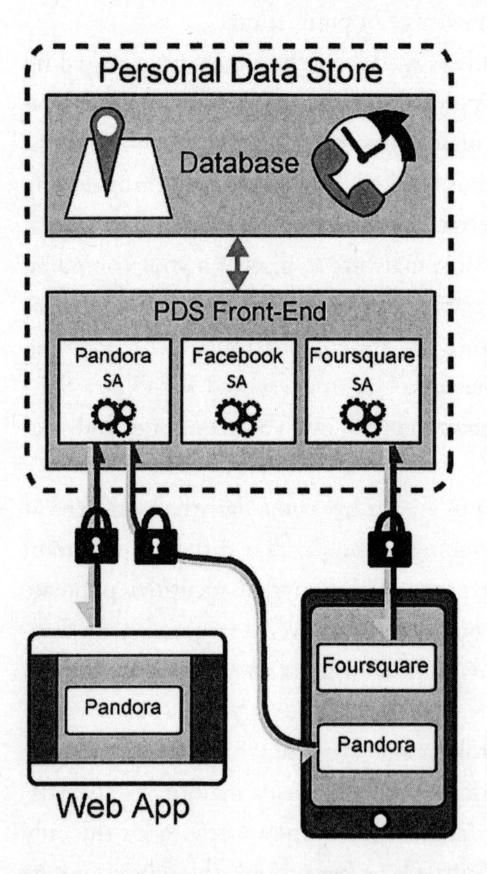

From a technical standpoint, the ability to let users share only the data they'd like to reveal in unique situations already exists. Sandy Pentland's Trento experiment,[17] for instance, utilizes software called openPDS ("personal data store"), which lets people control the flow of their data. This gets a bit geeky, but here's how openPDS works, as explained on its site in its "Only Answers, No Raw Data" section:

With SafeAnswers generic computations on user data are performed in the safe environment of the PDS, under the control of the user: the user does not have to hand data over to receive a service. Only the answers, summarized data, necessary to the app leave the boundaries of the user's PDS. Rather than exporting raw accelerometer or GPS data, it could be sufficient for an app to know if you're active or which general geographic zone you are currently in. Instead of sending raw accelerometer readings or GPS coordinates to the app owner's server to process, that computation can be done inside the user's PDS by the corresponding Q&A module.[18]

Translation: The raw data representing your key personal information is not required by apps or services to provide their core benefits. So analytics can be done inside or within your personal data store so no raw information is provided to outside parties. This limits the possibility of services giving this raw data to third parties with whom you have no previous relationship.

Many data banks or stores like this one also let you "kill" data, meaning the system will report any unauthorized access of information by outside parties and you can simply delete it, as if it were a file on your computer.

Facecloud is based on a system like openPDS. It provides a framework to implement VRM, or individually controlled flows of personal data. You may think it's too geeky to be applicable to your life, but companies such as Personal.com already feature services like Fill It,[19] which lets you enter your PII data and personal information one time to then be populated safely on any online forms you use in the future. The service is a form of a personal cloud and after you set up your preferences once, you don't have to reenter them. So ease of use for something like openPDS will be much greater than with the current way most of us do passwords or deal with our personal data.

From Vertical to Personal

The Facecloud idea goes beyond the framework for data or password exchange. As I outlined in the various scenarios around what information a person could see in a virtual environment, there is a contextual sharing aspect to personal data we'll experience in the near future. Whereas personal clouds or banks store and share data safely, the dashboards translating the data about our lives, called Life Management Platforms, will act as a kind of "Internet of Pings," or central database connecting all the signals surrounding our data into one platform we curate to organize our lives.

European analyst company KuppingerCole[20] has a report on this trend, "Life Management Platforms: Control and Privacy for Personal Data,"[21] that references Doc Searls and his ideas on VRM from his book *The Intention Economy.*[22] KuppingerCole feels these platforms will have a profound effect on the everyday life of people around the world within the next ten years. Here's an example of how the company feels a Life Management Platform would function in respect to your car:

> *Some years from now, my car will be accessible through a virtual key which is stored in my private domain, together with all information relevant for the usage and maintenance of that car. This will be kind of a digital driver's book, which would even report an engine fail to your garage if you wish it to do so (and only then). Thus, Life Management Platforms will become a key enabler for, among other things, the really connected car of the future.*[23]

This context of total life connection goes far beyond the simple sharing of data. These Life Management Platforms will let us utterly personalize the tiniest nuances of our everyday existence. For instance, people may install Breathalyzer technology in their cars to keep them from starting if the system recognizes alcohol on their breath. Or if there are multiple drivers in a family using the same car, each virtual key would be synced with their individual Pandora or Spotify accounts. Life Management Platforms will provide this same level of hyper-personalization for our health and medical bills, homes, insurance, and dating lives. Protected by a personal cloud architecture, these digital dashboards will transform how we see and experience the world.

"The HAT Project wants to answer, What's the value of data that allows us to make better decisions?" HAT stands for Hub of All Things,[24] a personal data platform that functions within a Life Management context. The platform allows individuals to buy apps that let them analyze or trade their data within a protected and relationship-based environment. I interviewed Irene Ng,[25] quoted above, professor of marketing and service systems, WMG, at the University of Warwick and the principal investigator for the project, which had its public release in 2015. During our chat, she pointed out that a primary difficulty most people have in understanding the value of their personal data is how it's verticalized by the industries who wish to mine and measure it. "We're forced into a data-collecting world that sits within silos—bank information, our blood pressure. So the data is formatted and collected according to these verticals."

But as Ng points out, humans don't think in verticalized ways about their lives. She eschews the notion of a "smart city" or "smart office," focusing instead on pragmatic implementations of data usage for individuals. For instance, she talks about all the data she aggregates in the time period between seven thirty and eight a.m., when she's planning her day, a time period she calls GRIM (Getting Ready in the Morning). This data includes water usage for her shower, the food she eats for breakfast, plus the clothes she wants to wear based on real-time weather conditions. Today this data may be collected by dozens of apps, which makes it cumbersome for individuals to manage. The HAT dashboard or a Life Management Platform does away with this clutter.

The platform also contextualizes security such as openPDS, providing "direct data debits" between two individuals or an individual and an outside organization. So when Irene's husband wants to access her location information, the couple has a direct data debit and information is exchanged solely between their two servers. There is no centralized manufacturer, or other data interloper like Google, to complicate transactions. And as a history of these data debits grows between individuals, the insights available about their lives will grow in complexity and value.

People will also be able to utilize this data debit mentality with advertisers and brands, but as the primary caretakers of their data, individuals will provide much more value to these relationships. Now the nuanced insights and analysis about their lives will be curated within the direct context of how or what they'd like to buy. It's a win-win for individuals and the companies open to transparent data relationships with their customers. The only losers are opaque data brokers or advertisers not willing to risk direct contact with the individuals they supposedly serve.

From Technical to Transformative

The current Internet economy isn't going anywhere soon. It makes a lot of people money, and most individuals don't recognize the value of their personal data, so they aren't upset at its theft. But the tide is turning. It began with Edward Snowden's revelations and grows with each new security scandal or data breach. On an individual level, people will tire of not being able to access all their medical and health records in one place. Or they'll want to block their health insurance companies from accessing their wearable health trackers to broadcast their data to their employers. Hopefully people won't be scared into using personal clouds or Life Management Platforms. Hopefully they'll simply realize the massive benefits, profits, and insights they'll gain if they do.

Once institutionalized, these platforms will provide frameworks for massive positive cultural change. Like the idea of community altruism I described in my opening story, a similar example was outlined by Irene Ng during our interview for the new types of social economies that will emerge from these platforms:

> *Imagine I have a cardboard box about four feet square in size. It's linked up to my HAT and equipped with sensors that transmit data to a local food bank. When I have food I know I'm not going to eat but is about to expire, I put it in that box. The food bank has servers in boxes like this throughout my neighborhood, and also knows which*

volunteers would be willing to gather those boxes in my community at any specific time. This means the HAT platform will have a huge collaborative consumption layer with much lower coordination costs than those that currently exist. This type of practice could bring about the rise of a whole new social economy.

A lot of the benefits of the New Deal on Data have to do with organization. In the same way you'd want to curate your own mind file, deciding which tweets, photos, and videos would represent your identity to the world, you should want to control your data. It's not an issue of privacy or philosophy as much as it is about controlling the assets that are rightfully yours. You put your money in a bank. Doesn't your digital identity deserve the same level of care?

Here are the primary ideas from this chapter:

- **Vendor relationship management.** VRM is growing in popularity, but faces a number of hurdles. Individuals don't understand the value of their data, and many advertisers and organizations feel CRM lets them keep the upper hand in relationships with their customers. Savvy companies, however, will understand individuals' control of their data means they'll be able to share deeper and richer context about their lives, thereby increasing opportunities for deeper relationships and purchase.
- **Personal clouds.** People aren't aware how much of their personal data is shared and sold by organizations they don't know. Data clouds allow individuals to be at the center of their own data universe, determining whom they want to share data with and under what circumstances. Clouds provide the only architecture for individuals to pursue in a digital economy versus the chaos that currently exists.
- **Life Management Platforms.** It took a while for websites to become easy to navigate for everyday visitors. User interface

and user experience became the way organizations communicated how they wanted people to navigate their sites in a simple and seamless fashion. Life Management Platforms will provide a similar ease of use with dashboards for our personal data within the context of our daily lives.

eight | A Vision for Values

Winter 2014

"Six laps," I pant as I hold out fingers on both hands, a reminder of how many times I've run around my local outdoor track. I used to mix up how many laps I'd run in the past, so this physical act helps me remember. Three times around the track is a mile, so within twentyish minutes I'd hit the two-mile mark. I've been actively running for more than four months. When I began, I could barely run two laps without needing to stop and walk. Now most of the time I only stop because I have to get back to work or be home for the kids after school.

I wear a Withings fitness monitor[1] clipped to my shorts when I run. A black wearable device about the size of two quarters put together, it measures my steps, heart rate, and sleep. It syncs with an app via Bluetooth so I can measure my health data, which I've done for the past few months. My daily goal is ten thousand steps, which is about four or five miles. I've only missed making that goal in three days over the course of four months, due to Thanksgiving and a few days when I was sick. And now I crave my runs or workouts at the gym. I savor the ritual of watching my health app's progress bar click past ten thousand steps at night, the digital readout displaying, "You made your daily steps goal!" If I don't get ten thousand steps, I walk in place watching *Game of Thrones* or a different show, much to the bemusement of my kids. I don't mind, because I committed to the daily steps. Beyond getting fit, I'm achieving a predetermined daily goal.

I'm a musician. I play blues harmonica and guitar and sing in a few local bands when we can get gigs. We do covers of B.B. King, Stevie Ray Vaughan, and whatever assortment of tunes the guys in the band feel like playing. I'm blessed to live so close to New York City, because the guys I play with are hard-core working musicians. One of our regular drummers even tours with the original Blues Brothers band. I mention the music thing because when I run I get to listen to bands I normally don't take time for when I'm not exercising. Delbert McClinton is particularly suited for running. I play a few of his songs at specific points in my run when I start to lag. Then during my sprints I play Led Zeppelin's "Rock and Roll." It may be cliché, but it's nearly impossible not to sprint when I hear Bonzo's opening drum riffs.

Other times I listen to NPR or podcasts while I exercise. I find this is most helpful in the middle of my hour-long aerobic workout, as I can forget about the rhythm of my feet and focus on the content of the show. Sometimes I feel pretentious listening to NPR, wondering if I'm elitist to dedicate brain space to listen to a program about animals liking classical music. But then I just ignore myself and click on content about something I want to learn about. I would rue the day I'd ever not want to learn something new.

My maternal grandfather was a grade school principal for more than fifty years, and lived to be ninety-five. I never visited him when he wasn't fascinated by an article in one of his magazines, or a subject he'd learned about from watching *60 Minutes*. The image of him pointing to a blackened velvet sky one summer night when I was a boy, teaching me the names of the constellations, is vivid in my mind. Surrounded by the thrum of crickets, my grandfather wasn't just teaching me facts about stars. He was imbuing me with his insatiable curiosity and his need to share what he knew. He was a quiet man, but a joyful one. His eyes often twinkled behind his thick glasses, his hand reaching to rub his forehead to rearrange the thoughts contained within.

Last August, on our wedding anniversary, I told my wife I was going to get in shape. We're both middle aged and are getting to that era in life when we begin to hear about friends dying or becoming gravely ill. The events serve as painful reminders of our mortality, and persuaded me to reflect on areas of my life I'd avoided before. For one thing, I'd gotten grossly overweight while working from home the past number of years, and wanted to show Barbara I loved her by giving her a gift that demanded repeated action to achieve success. Her love language was acts of love. Being a writer, words are my love language. It took us a number of years of our marriage to recognize that you need to love your spouse in the language they use versus your own.

So I got the Withings monitor and started to run. Now almost four months later I've lost more than thirty pounds. My face has definition again. The new "thin pants" I bought a month ago are already too big for me. I'm lifting weights and actually have biceps that don't jiggle when I move. I'm at the gym every day and even get man nods of approval from a few of the regulars who can bench more than I weigh. It feels awesome to gain their approval. It's not because I'm an enviable male specimen, mind you (at least not yet). I think they just appreciate I show up every day to sweat my butt off. Literally.

I don't know how much more weight I'll lose. My new goal is to lose forty-six pounds by my forty-sixth birthday. That seems like a cool aspiration, even T-shirt worthy. I'm not sure if I'll be able to keep the weight off. But I've essentially reset my metabolism. I've also made exercise a genuine habit, and am eating a lot better. Most important, I've done my best to show Barbara she has one less person in her life to worry about, healthwise. I can still get hit by a bus. But now I have a better chance of outrunning it.

I savor working out at my gym. I like seeing familiar faces. I have my favorite treadmill and elliptical machines. I have favorite routines to work out various muscle groups, and know when to push myself versus back off. The rituals have become comforting. The fitness extends beyond my physical self to my mental and emotional health as well. I can't keep negative emotions from showing up, but I can manage them by exercising.

Running outdoors provides an even deeper sense of meaning for me. A lot of it is the visceral experience—wind, warm or biting cold, as it tries to steal the baseball cap off my head. The ground, hard and uneven, the bumps and fissures mapped in my brain over time, increasing my confidence as I wend my way around the track.

Plus, there's the prayer. Once a run, I try turning off my music to center myself. I listen to myself breathe, take in my natural surroundings. Then I pray. Other people meditate—I supplicate. Running is often when I feel closest to God.

People may not like me, my books may never sell, I may never have enough money. These are the worldly mantras I push away as I run, allowing deeper truths to emerge as sweat soaks through my shirt. Then, for a brief yet tantalizing moment, I stop feeling my body move. I stop thinking about thinking. I'm just in the moment. I said I was going to get in shape, and I'm in that process. I don't know the future, but for now, I am content.

I'll let tomorrow take care of itself.

Markets and Morals

When your values are clear to you, making decisions becomes easier.[2]

—ROY E. DISNEY

I don't hate corporations, and I don't hate advertising when it focuses on solid storytelling and permission-based marketing versus surveillance. What I detest is the idea of basing someone's worth on how much money he or she makes. Similarly, I loathe the notion that a country's well-being is determined by its GDP, or other financial metrics. These two ideas mean children all over the world grow up valuing dollars over sense of purpose.

We care about the things we count. Money, besides being something we need in order to live, is easy to count. That's a great quality when you (a) have money to count in the first place, (b) are good with numbers, and (c) are part of a culture that values the accumulation of wealth. But the esteem we attribute to wealthy people doesn't necessarily reflect their positive character. And the amount of time we spend pursuing money doesn't correlate with an increase in well-being.

My point is not that money is the root of all evil. But far too often it's the route to identity. If you're rich and horrible, people often still respect you. If you're poor and brilliant you're typically shunned. The fact that this is the way of the world does not make it correct. It means the world has been trained to value money and those who have it over other, harder-to-identify characteristics.

Until now.

Now we have influencers—people whose social media channels have millions of views because of their creativity or, in some cases, their snarky points of view. Their popularity may bring them riches, but they originally captured attention because their work moved people.

Now we also have wearable devices measuring our actions, broadcasting our accountability. Did we go to the gym today? Get the sleep we said we should? This accountability will begin to influence how others think about us in new ways. People at your office may start to wonder why you've bought three different exercise devices but haven't gotten in shape. How would this precedent relate to your potential productivity at work? Soon, people at companies who opt out of wearable fitness programs will be looked down upon. Why won't Karen use a step counter? Doesn't she know her choice to be heavy means our insurance rates stay high? As the Internet of Things matures, we'll be tracking our values as much as we're tracking our data.

In Austin, Texas, multiple families have been demonstrating this type of behavior in Mueller, a planned green community tracking electricity usage. As Bryan Walsh reports in his June 2014 article for *Time* magazine, "Is This America's Smartest City?" locals are learning in real time how their connected appliances and electric cars affect their bills and the local electric grid.[3] It's a perfect example of how our actions affect not only our own lives but also those of our surrounding community. In the case of Mueller, one example of neighborhood awareness comes when families choose to charge their electric cars. While utilities felt most people would opt to charge after work, taxing local electric grids, families have actually opted for overnight charging during off-peak, cheaper hours. Engineers are also testing ways for electric cars to store enough excess solar power during the day to power homes at night. As Jim Robertson, one of the participants in the project, notes in the article, "It really shows the value of having a smart home."[4]

But the value of having a smart home goes beyond money for these families. They're also conserving the environment, putting less strain on the community, and reducing waste. These are all values that can be tracked, and are being measured before and after their implementation. There is the possibility that neighbors not participating in the program will get judged negatively. Why are the Joneses running their air conditioner at all hours and taxing the neighborhood? These are some of the cultural paradigms we're going to face, in which money still plays a role but values are bigger drivers of behavior. People will define and broadcast character like never before.

Values in Full View

The story opening this chapter is true, although my wife isn't named Barbara. As I sit writing this book, I've lost thirty-one pounds over the course of four months. And my goal is to lose forty-six pounds by my next birthday. While I was inspired to start this process for my wife and kids, to minimize their stress at my girth, I also realized that by not spending any time exercising, I wasn't being true to my beliefs.

I learned this by tracking my values. Literally.

I got the idea from a friend of mine named Konstantin Augemberg.[5] He called his study "Hacking Happiness"[6] (he knew that, at the time, I was writing a book by the same title, so I had a special affinity for his project right away) and used a few well-established theories to track how his actions reflected his stated values. First off, he created a daily values survey by using the rTracker app,[7] which lets you easily set up whatever metrics you want to measure for a certain activity. Then he included metrics from two well-known sources regarding values tracking, the

Ryff Scales of Psychological Well-being[8] and the Schwartz value theory.[9] For all of these elements, he wanted to test what he called his "values dissonance theory"—the idea that "not being able to live according to your values causes unhappiness."[10] This concept made enormous sense to me, and merited testing in my own life. Here are the fourteen elements Konstantin tracked for his values dissonance theory, along with a diagram showing how he measured them via the app he created with rTracker:

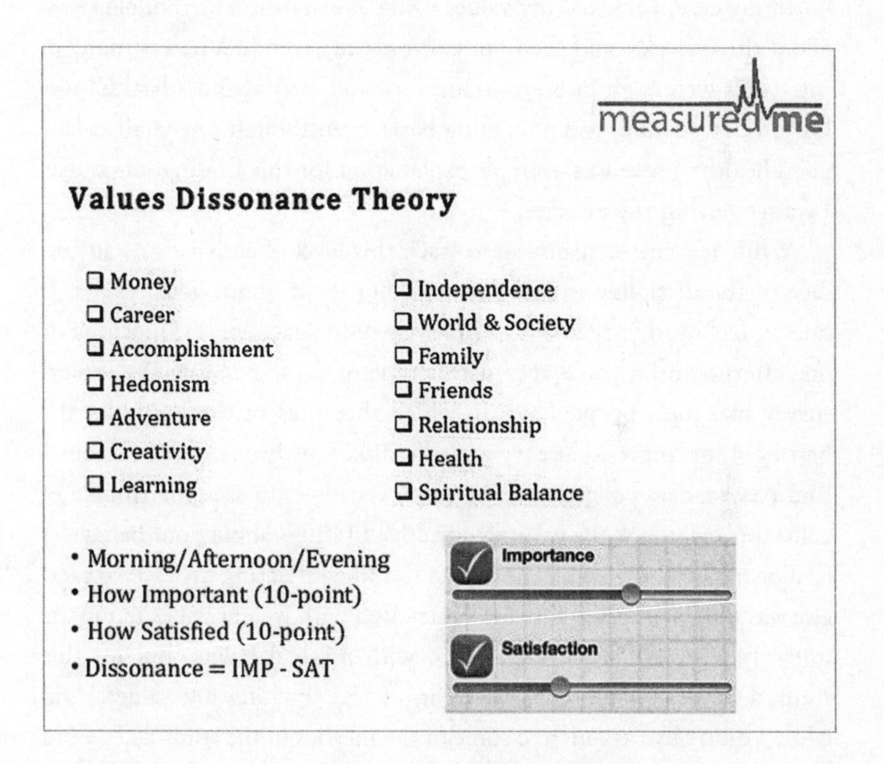

How this process worked was that Konstantin was prompted three times a day via his app to measure two criteria (importance and satisfaction) for each of the fourteen metrics he was tracking. He did this for thirty-four days, discovering various insights—for instance, spending time with his partner or dog, cooking, and relaxing helped him be happier. His results also showed spending time at work and on fitness were negatively associated with happiness. From a pragmatic standpoint, Konstantin told me he used his tracking metrics to ask his boss about changing his role at work to better suit his values as he'd measured them in his test. His boss was impressed with this proactive stance, and when Konstantin grew more productive and happy at work they both celebrated his success. Seen in this light, tracking values is not just an academic exercise, but a tool to identify and pursue activities that can increase your well-being.

In my case, I tracked my values using Konstantin's methodology for about three weeks, and then noticed a glaring trend. Whereas many of my scores were high in both importance and satisfaction, "health" was low on both counts on an ongoing basis, even though I generally value good health. There was a simple explanation for this finding, of course: I wasn't getting any exercise.

While it's time-consuming to track this level of activity, one advantage of the discipline is that you tend not to lie about your results. It doesn't feel overly negative to admit you didn't exercise one morning or one afternoon. So you'll record this type of data, avoiding the typical survey bias most people have, in which they present themselves in the best light, or forget what they actually did. This is one of the reasons I find passive data collection to be so powerful—you save the trouble of constant tracking while gaining objective findings about your behavior.

For me, seeing a graph showing a flat line reflecting my lack of exercise was quite compelling. I've dealt enough with weight issues in my life to easily feel guilt when confronted with my bad habits. But for this study, I wasn't trying to lose weight—I was tracking my values. And while I didn't live as much to some of the metrics in the study as I would have liked (e.g., creativity and learning), I didn't spend *any* reported time on my health. As a person of integrity, how could I feel good about myself if there was, as Konstantin called it, such a dissonance between what I told myself was important in life and what I acted out every day?

This knowledge, while not fun to discover at the outset, has been extremely empowering for me. Now, instead of just wanting to fit into my thin jeans, I frame going to the gym or running as a manifestation of my values. I try to maintain a balance in my quest, as equilibrium among metrics such as Konstantin's tends to bring the greatest well-being over time. However, for this era in my life I'm also making up for missed health opportunities in the past. While I can't go back and regain the personal integrity I sacrificed during those times, I'm inspired to maintain my daily regimen because I want to live by my values. I want to be a man of my word—to my family, to the world, and especially to myself.

Carol D. Ryff, in her 1989 paper, "Happiness Is Everything, or Is It? Explorations on the Meaning of Psychological Well-being,"[11] created what are called the Ryff Scales of Psychological Well-being,[12] mentioned as part of Konstantin's work. A full inventory, or survey, created around the work comes in both a long and medium format that reflect six areas of psychological well-being. People taking the survey rate answers on a scale of 1 to 6, where 1 indicates strong disagreement. Here are some example statements from the Ryff inventory as written by Tricia A. Seifert, from the University of Iowa.[13] Go through them and rate yourself from 1 (not at all) to 6 (completely) as a way to get a grasp on the state of your well-being:

The following are example statements from each of the areas of well-being measured by the Ryff inventory:

- **Autonomy**
    - I have confidence in my opinions, even if they are contrary to the general consensus.
- **Environmental Mastery**
    - In general, I feel I am in charge of the situation in which I live.
- **Personal Growth**
    - I think it is important to have new experiences that challenge how I think about myself and the world.
- **Positive Relations with Others**
    - People would describe me as a giving person, willing to share my time with others.

- **Purpose in Life**
  - ✧ Some people wander aimlessly through life, but I am not one of them.
- **Self-Acceptance**
  - ✧ I like most aspects of my personality.[14]

If you're like most people, you've never been exposed to these types of questions before. Aren't they enlightening? Can you start to get a sense of how actionable each of these areas could be if you began to track them? Part of Ryff's contribution to positive psychology is reflected in the inclusion of her work in this field. As she points out in her paper, "Increased interest in the study of psychological well-being follows from the recognition that the field of psychology, since its inception, has devoted much more attention to human unhappiness and suffering than to the causes and consequences of positive functioning."[15] We'll discuss more about positive psychology in a later chapter. But its main premise is that people can engage in regular, measurable behaviors that will increase their happiness, like a gym regimen for well-being.

This is a massive revelation. While huge amounts of empirical study have been devoted to depression or psychosis, it's only in the past fifteen years or so that the field of positive psychology has worked to identify the behaviors manifesting in happiness. It makes sense, then, that so many of us rely on knowing if or when we're happy based on mood. But it's impossible to pursue happiness for its own sake—happiness or the lack thereof is what *arrives in the process*.

Shalom H. Schwartz, in his seminal paper, "Universals in the Content and Structure of Values: Theoretical Advances and Empirical Tests in 20 Countries,"[16] identified ten common values held by study participants in forty postindustrial countries.[17]

While it's always challenging to assume a set of universal values for any individual or set of citizens, Schwartz's work finds that people of different backgrounds typically share a core set of morally based beliefs. They are not necessarily moral absolutes, because the manner in which different people or countries interpret the values determines their behaviors. But identifying and naming these values gives us a consistent vocabulary and set of criteria we can use to measure what counts in our lives on a deeper level than money.

If we weren't all defined as consumers, personalization algorithms could be focused on what brings us well-being versus influencing what we buy. If increased productivity and profits weren't the primary values we measure globally because of the influence of GDP, we'd be free to test and measure other metrics that could make us happier. If we're at a time in history when neuroscientists and AI experts are considering how human consciousness can be replicated in silicon, can't we update an economic structure such as the GDP—which was created before the Second World War? And as hard as it will be to define ethical standards around the creation and proliferation of AI, shouldn't that be our focus *before* or at least *during* its manufacture?

Case Study: Tracking Values and the H(app)athon Project

I founded an organization called the H(app)athon Project[18] in 2012 to study the relationship between emerging technology and well-being. In 2014, I created a study for tracking values with one of our board members, Peggy Kern.[19] Dr. Kern is a senior lecturer at the Melbourne Graduate School of Education, in Australia. She is a thought leader in the field of positive psychology, and spent a good portion of her career working directly with Martin Seligman[20] at the University of Pennsylvania, whom many call the father of positive psychology.

Peggy and I became interested in the relationship between values and well-being and wanted to emulate Konstantin's app for the general public. While we originally designed the survey to be taken over the course of three weeks, during which daily e-mail prompts would ask if you were living to your values, I've included the core of our work below for you to test your values *right now*. The Pre-Tracking Well-being Assessment lets you rate your baseline well-being and values, indicating what areas are most important to you. After this, try filling out the values form at the end of the day for a number of days, rating yourself on a 1-to-10 scale of whether or not you feel you lived according to your values that day. (I've also included this tool at the end of the book, at heartificial intelligence .com, and on my website, johnchavens.com.)[21]

Connecting Happiness to Action

PRE-TRACKING WELL-BEING ASSESSMENT

In his 2011 book, *Flourish*,[22] Dr. Martin Seligman, distinguished professor of psychology at the University of Pennsylvania and founder of the field of positive psychology, defined five pillars of well-being, called PERMA (Positive emotion, Engagement, Relationships, Meaning, and Accomplishment). The PERMA-Profiler measures these five pillars, along with negative emotion and health.[23]

Please read each of the following questions and then select the point on the scale that you feel best describes you. Please be honest—there are no right or wrong answers. 1 indicates "not at all" or "never," while 10 indicates "completely" or "always."

In general, to what extent do you lead a purposeful and meaningful life?	1 2 3 4 5 6 7 8 9 10
How much of the time do you feel you are making progress towards accomplishing your goals?	1 2 3 4 5 6 7 8 9 10
How often do you become absorbed in what you are doing?	1 2 3 4 5 6 7 8 9 10
In general, how would you say your health is?	1 2 3 4 5 6 7 8 9 10
In general, how often do you feel joyful?	1 2 3 4 5 6 7 8 9 10
To what extent do you receive help and support from others when you need it?	1 2 3 4 5 6 7 8 9 10

In general, how often do you feel anxious?	1 2 3 4 5 6 7 8 9 10
How often do you achieve the important goals you have set for yourself?	1 2 3 4 5 6 7 8 9 10
In general, to what extent do you feel that what you do in your life is valuable and worthwhile?	1 2 3 4 5 6 7 8 9 10
In general, how often do you feel positive?	1 2 3 4 5 6 7 8 9 10
In general, to what extent do you feel excited and interested in things?	1 2 3 4 5 6 7 8 9 10
How lonely do you feel in your daily life?	1 2 3 4 5 6 7 8 9 10
How satisfied are you with your current physical health?	1 2 3 4 5 6 7 8 9 10
In general, how often do you feel angry?	1 2 3 4 5 6 7 8 9 10
To what extent do you feel loved?	1 2 3 4 5 6 7 8 9 10
How often are you able to handle your responsibilities?	1 2 3 4 5 6 7 8 9 10
To what extent do you feel you have a sense of direction in your life?	1 2 3 4 5 6 7 8 9 10
Compared to others of your same age and sex, how is your health?	1 2 3 4 5 6 7 8 9 10
How satisfied are you with your personal relationships?	1 2 3 4 5 6 7 8 9 10
In general, how often do you feel sad?	1 2 3 4 5 6 7 8 9 10
How often do you lose track of time while doing something you enjoy?	1 2 3 4 5 6 7 8 9 10
In general, to what extent do you feel contented?	1 2 3 4 5 6 7 8 9 10
Taking all things together, how happy would you say you are?	1 2 3 4 5 6 7 8 9 10

Values

Scientific research shows that when we don't live in accordance with our values, our well-being decreases. It is also the balance regarding the interplay of our values that dictates many of the actions we take in our lives.

Please take a moment to think about who you are and what you value in life. Then read the following descriptions of different people. For each one, read the description, and then indicate how much the person in the description is like you. Be honest—there are no right or wrong responses. None of these are good or bad; they are simply descriptions of different people. For each of the following, indicate how much the person in the description is like you (1 = not at all like you, 10 = completely like you).

VALUE AND DESCRIPTION	RATINGS
WORK: This person enjoys working hard, finding a lot of meaning in the daily activities, whether for paid employment or unpaid activities.	1 2 3 4 5 6 7 8 9 10
TIME BALANCE: This person enjoys keeping a balance between work, family, and social aspects of life, allowing for time for excitement, rest, and stimulation.	1 2 3 4 5 6 7 8 9 10
EDUCATION, ARTS & CULTURE: This person enjoys learning. He or she likes to visit museums and other cultural centers, or engage in artistic pursuits.	1 2 3 4 5 6 7 8 9 10
ACHIEVEMENT: This person likes to have people recognize his or her achievements. Being very successful is important to this person.	1 2 3 4 5 6 7 8 9 10

MATERIAL WELL-BEING: This person likes to have a lot of money and expensive things. It is important for this person to be rich.	1 2 3 4 5 6 7 8 9 10
HEALTH: This person likes to engage in healthy behaviors. Staying physically or mentally fit is important to this person.	1 2 3 4 5 6 7 8 9 10
GOOD TIMES: This person likes to have a good time, doing things that make him or her feel good throughout the day.	1 2 3 4 5 6 7 8 9 10
HELPING OTHERS: This person likes to care for and help others.	1 2 3 4 5 6 7 8 9 10
SECURITY: This person likes to avoid anything that might be dangerous. It is important to live in secure surroundings and feel safe.	1 2 3 4 5 6 7 8 9 10
NATURE: This person likes to spend time in nature. He or she seeks out green spaces, and strives to care for natural resources.	1 2 3 4 5 6 7 8 9 10
FAMILY: This person likes to spend time with his or her family. Filling family needs is important to this person.	1 2 3 4 5 6 7 8 9 10
SPIRITUALITY: This person feels connected to something higher than him- or herself. Feelings of spirituality or religious or spiritual practices are important to this person.	1 2 3 4 5 6 7 8 9 10
OTHER values not listed here:	1 2 3 4 5 6 7 8 9 10

For further testing, go to http://www.yourmorals.org/explore.php and click on the Register link next to "Schwartz Values Scale."

Take a minute and look at your results. How high or low did you score on the different well-being pillars? Which values did you rate highly and which ones were lower?

The next step is to track these values for a few days, rating yourself on a 1-to-10 scale of whether or not you feel you lived according to your values that day. (I've also included this tool at the end of the book and on my website.)

VALUE	DAY 1	DAY 2	DAY 3	DAY 4	DAY 5
Work	1 2 3 4 5 6 7 8 9 10	1 2 3 4 5 6 7 8 9 10	1 2 3 4 5 6 7 8 9 10	1 2 3 4 5 6 7 8 9 10	1 2 3 4 5 6 7 8 9 10
Time balance	1 2 3 4 5 6 7 8 9 10	1 2 3 4 5 6 7 8 9 10	1 2 3 4 5 6 7 8 9 10	1 2 3 4 5 6 7 8 9 10	1 2 3 4 5 6 7 8 9 10
Education, arts & culture	1 2 3 4 5 6 7 8 9 10	1 2 3 4 5 6 7 8 9 10	1 2 3 4 5 6 7 8 9 10	1 2 3 4 5 6 7 8 9 10	1 2 3 4 5 6 7 8 9 10
Achievement	1 2 3 4 5 6 7 8 9 10	1 2 3 4 5 6 7 8 9 10	1 2 3 4 5 6 7 8 9 10	1 2 3 4 5 6 7 8 9 10	1 2 3 4 5 6 7 8 9 10
Material well-being	1 2 3 4 5 6 7 8 9 10	1 2 3 4 5 6 7 8 9 10	1 2 3 4 5 6 7 8 9 10	1 2 3 4 5 6 7 8 9 10	1 2 3 4 5 6 7 8 9 10
Health	1 2 3 4 5 6 7 8 9 10	1 2 3 4 5 6 7 8 9 10	1 2 3 4 5 6 7 8 9 10	1 2 3 4 5 6 7 8 9 10	1 2 3 4 5 6 7 8 9 10
Good times	1 2 3 4 5 6 7 8 9 10	1 2 3 4 5 6 7 8 9 10	1 2 3 4 5 6 7 8 9 10	1 2 3 4 5 6 7 8 9 10	1 2 3 4 5 6 7 8 9 10
Helping others	1 2 3 4 5 6 7 8 9 10	1 2 3 4 5 6 7 8 9 10	1 2 3 4 5 6 7 8 9 10	1 2 3 4 5 6 7 8 9 10	1 2 3 4 5 6 7 8 9 10
Security	1 2 3 4 5 6 7 8 9 10	1 2 3 4 5 6 7 8 9 10	1 2 3 4 5 6 7 8 9 10	1 2 3 4 5 6 7 8 9 10	1 2 3 4 5 6 7 8 9 10
Nature	1 2 3 4 5 6 7 8 9 10	1 2 3 4 5 6 7 8 9 10	1 2 3 4 5 6 7 8 9 10	1 2 3 4 5 6 7 8 9 10	1 2 3 4 5 6 7 8 9 10
Family	1 2 3 4 5 6 7 8 9 10	1 2 3 4 5 6 7 8 9 10	1 2 3 4 5 6 7 8 9 10	1 2 3 4 5 6 7 8 9 10	1 2 3 4 5 6 7 8 9 10
Spiritualty	1 2 3 4 5 6 7 8 9 10	1 2 3 4 5 6 7 8 9 10	1 2 3 4 5 6 7 8 9 10	1 2 3 4 5 6 7 8 9 10	1 2 3 4 5 6 7 8 9 10
Other	1 2 3 4 5 6 7 8 9 10	1 2 3 4 5 6 7 8 9 10	1 2 3 4 5 6 7 8 9 10	1 2 3 4 5 6 7 8 9 10	1 2 3 4 5 6 7 8 9 10

Peggy and I created blog posts for survey participants[22] who tracked their values in our experiment. I've included content from those posts here so you can measure the effects of your own tracking.

General Insights (Well-being)

As we noted in our survey, we believe that if you want your life to count, you have to take a count of your life. That's what you've done by taking a measure of your well-being and your values. When looking at your well-being and values scores over the past few days/weeks, try to not be self-critical but simply think about what insights come to you as you study the results.

For instance, for your well-being scores, if you had higher numbers upon completing the survey than when you started, the act of measuring yourself may have been a positive experience. If your scores were lower upon completing the survey, you may have been in lower spirits at the time you filled out your answers, or measuring yourself may have been a difficult experience. While we sincerely hope this survey will help you increase your well-being, sometimes drops in well-being can help you identify areas to improve. So we encourage you to focus more on analyzing *why* your scores went up or down. Did you have a particularly interesting or challenging couple of weeks? Are there behaviors you can pinpoint that helped you increase your well-being and that you should continue doing, or behaviors that lowered your well-being that you can avoid?

General Insights (Values)

This same logic applies to your values. We designed our questions so that your answers could provide you insights that might be actionable in some way. Along those lines, when looking at your values upon completing your tracking, ask yourself the following questions:

- Did identifying my values help me understand them better?
- Did rating my values change my perception of what I truly hold dear?

- Did tracking my values bring insights into how I spend my time in relation to what I think I care about?

There are no right or wrong answers for these questions—they're designed to help you genuinely examine your values as you try to live them out each day. We do hope, however, that your insights regarding your values prove to be *actionable*. Along those lines, here are some questions that we hope prove helpful:

- Why do you think certain value categories (work, family) were so much higher when you first filled out the survey than when you ended the survey?
- Do you think tracking your values helped you realize which ones you're actually living according to each day?
- Do you think your results have shown you one or two (or more) value categories where you can spend more or less time to find more balance or well-being in your life?
- What keeps you from trying to live according to your values? Are those things you can minimize in an effort to increase your well-being and happiness?

The image below illustrates how you and others define well-being:

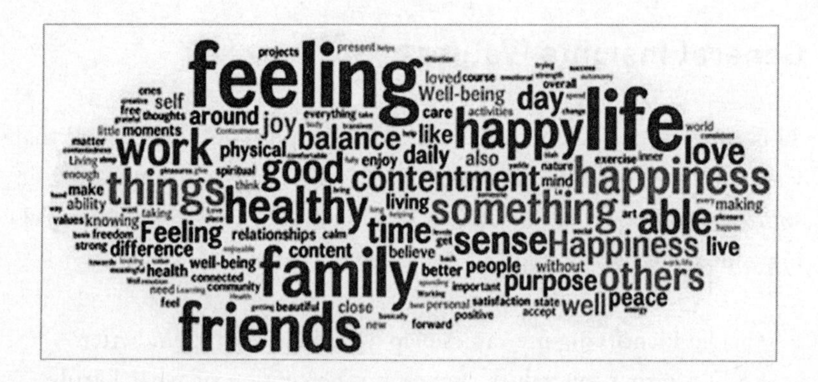

Below are two graphs showing a composite of the first forty-seven people who completed the values survey. Here is a graph showing aggregate results for well-being:

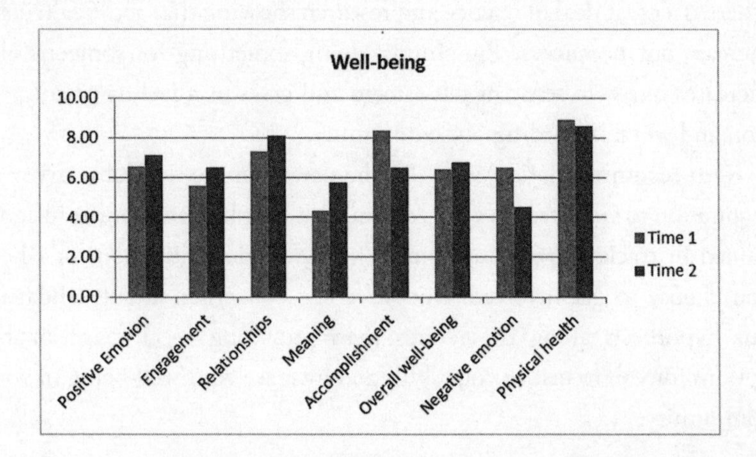

Overall, people increased well-being scores and decreased their negative emotions. Our hope is that the act of identifying and tracking their values contributed to these results.

The importance of values generally increased over the two-week period, as shown in the following graph. Our hope is that differences are due to changing perceptions as a result of tracking values during that time. You'll also note that "work" decreased where "health" increased, demonstrating that tracking hopefully helped users balance their values better.

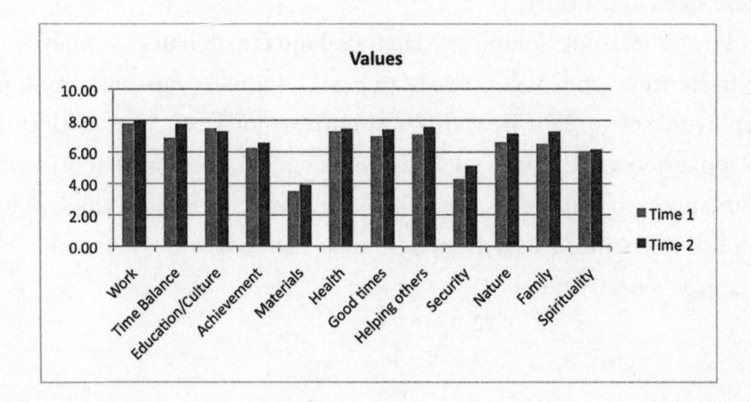

What Next?

There is a great deal of science and research showing that acts of altruism increase our happiness. Put simply, doing something for someone else increases our own sense of self-esteem and gives us a feeling of connection and worth regarding our community.

Our recommendation is to use the data from your values survey as inspiration to volunteer in your community. Our hope is that the insights gained in tracking your values may help provide a direction for where you choose to get involved. But we're less concerned about validating our hypothesis along these lines than providing specific, actionable options for you to utilize your skills and increase your well-being in your community.

Connecting Happiness to Action

Our hypothesis (and hope) is if you choose a volunteer area based on your values (in which your results show that living to those values successfully increased your happiness), your happiness or well-being should increase.

If you'd like to connect your values to volunteering, here are a few more steps to consider.

First, check out Volunteer Match (volunteermatch.org). Simply enter your location and Volunteer Match will provide you with multiple opportunities to help in your community—for free. After taking the H(app)athon survey, you can volunteer based not only on what organizations appeal to your heart, but which activities reflect your values. Below is a list of Volunteer Match interest areas matched with values from the H(app)athon survey.

Volunteer Match Cause Areas	Matching Values H(app)athon Survey
Advocacy	Achievement
Animals	Nature
Arts & culture	Education, art, culture
Board development	Achievement
Children & youth	Education, art, culture
Community	Family
Emergency & safety	Security
Employment	Achievement
Environment	Nature
Faith-based	Spirituality
Health & medicine	Health
Homeless & housing, hunger	Helping others
Justice & legal	Security
Politics	Achievement
Seniors	Family
Sports & recreation	Health
Women	Family

What do you value in life? Can you codify and test all those things? Odds are you know exactly how much is in your bank account right now. But aren't you also worth the effort of understanding what core values determine your well-being on a daily basis?

We shouldn't enter a future where advertising-driven algorithms determine the values in our lives. This is the worst form of artifice in terms of happiness, as the science of positive psychology shows the mere accumulation of stuff doesn't increase well-being but, in fact, hinders it. It's the AI version of the hedonic treadmill.

Beyond avoiding the existential threats strong AI may bring, we need to codify values such as the ones I hope you've tracked using the tools included in this chapter. This will allow us to name, validate, and scale the human traits that we care about most within the systems we've already begun creating. Recognizing these values will also help us set limits on automation, since we'll know which areas of our lives we don't want something else to take responsibility for.

Here are the primary ideas from this chapter:

- **Accountability goes public.** We're entering the era of the Internet of Things, in which the objects surrounding us will reflect and report on our actions more than ever before. Now character traits beyond the accumulation of money will be visible in real and virtual worlds, helping us define economies in ways that more accurately reflect well-being than wealth.
- **It's time we focused on the positive.** The empirical study of positive emotions, character, and strengths in psychology is a relatively new yet very powerful phenomenon. Positive psychology will hopefully soon become a common part of every person's health regime.
- **Make your values count.** If values are what guide our lives, shouldn't we be able to identify them? Once we do so, we can track them in the same way we measure the money we spend. Imagine how your world might change if the ledger for your life showed a surplus of well-being from actively pursuing your values versus just trying to widen your wallet.

nine | Mandating Morals

Summer 2037

"We're designing this guy to be a kitchen assistant for seniors living in people's homes. His nickname is Spat, short for *spatula*, since he comes preprogrammed to cook about twelve hundred types of meals." My son, Richard, paused and looked at me, smiling. "And yes, he cooks bacon. Crispy or non-crispy. Even soy."

I winced. "They have *soy* bacon? What's the point? Isn't it just salmon-colored tofu?"

Richard shrugged, his broad shoulders lifting his white lab coat. Recently he'd grown a full beard, and was keeping his typically unruly mane of blond hair neatly trimmed. He was a good-looking kid. "To each their own. I agree it's a weird placebo or substitution or whatever. But the vegan market is huge, and Spat aims to please." Richard nodded toward a five-foot robot within the glass-enclosed kitchen in front of where we were standing. Spat's face appeared cartoonish and friendly, his pear-shaped body giving the impression of a chubby, amiable chef. The kitchen was one of a half dozen testing areas where Richard and his colleagues at Moralign installed human values protocols into robots.

As artificial intelligence began to rapidly proliferate in 2015, programmers recognized the need to build uniform ethical codes into the algorithms that powered their machines. However, creating these codes proved to be an enormous challenge. The AI industry was vast, comprising multiple silos. Groundbreaking work on consciousness and deep learning[1] took place in academia, funded by outside corporations' research and development budgets. Many times these academics weren't aware of how their ideas would be translated into consumer products, which made ethical standards impossible to implement. Similarly, once products were developed they were handed off to lawyers struggling to create standards in a society where laws for robots didn't yet exist. While it was an exciting era to work in the legal profession, it was also a daunting one—until the turn of the century, all existing laws had been written by and for humans. Since technology was advancing faster than regulators could keep up, this also meant most laws were seen as restrictive toward innovation.

In 2017, Moralign was founded to tackle the issue of ethical standards from a new angle for the industry. Instead of requiring every AI manufacturer to hire ethical experts or lawyers to comply with regulations that hadn't been invented, Moralign proposed a different solution: We'll imbue your existing machines with human values programming and ethically test your products before releasing them to the public. This meant Moralign would be able to iterate their human ethics software while creating uniform industry standards based on marketplace testing. The company would serve as both a consumer protection agency and an R&D lab for top AI manufacturers around the world.

As a start-up, Moralign had twenty-two employees, comprising data scientists, neuroscientists, data programmers, sociologists, marketers, and actors. Like actors hired to train medical professionals, Moralign's performers worked with AI companies to write scripts designed to emulate the lives of the people who would be buying their products. Then the actors improvised common scenarios utilizing AI products within the confines of Moralign's laboratories. This allowed autonomous machines outfitted with myriad sensors to observe humans at work, home, or play. Initially Moralign had rented apartments and offices to test robots in these settings with their actors. Now sets were built in the company's spacious warehouse, making it feel like the back lot of a Hollywood film studio.

Richard was in charge of the acting program. He'd started off as a performer for the company after college, auditioning for Moralign's improvisational troupe when a friend's father introduced him to one of the company's founders. While actors sometimes had specific scripts that included language and taxonomies relevant to whatever products were being tested, more often performers were simply given scenarios and character descriptions to follow for their work. For Richard's audition, he'd been provided with case study material for a returning veteran with post-traumatic stress disorder (PTSD). Congress had provided funding for a leading MIT firm specializing in deep learning to create robot assistants that could help soldiers reacclimate to civilian life after being in combat. The company was going to be working with Moralign to implement its ethical protocols, so it was a perfect time for Richard to test his skills.

I'd helped Richard rehearse for the audition, which was a real joy because I'd been a professional actor. We watched a ton of movies and documentaries about combat and PTSD, and had lunch with friends of mine who created apps and tech for the military. Richard soaked everything up like a sponge and when he gave his performance for the Moralign team, even I forgot he was my son. He transformed into a different, but genuine, human being before our eyes. Applicants auditioned behind two-way glass like they have in cop shows, as they had to be comfortable being watched by Moralign staff and clients in their work. Had Richard been able to see my reaction to his audition, my tears of pride probably would have embarrassed him.

Along with his performing skills, Richard was great with clients. He could communicate as easily with geeky programmers as he could with chief marketing officers and patent lawyers. Deeply emotionally intelligent, he made anyone he was speaking to feel comfortable and heard. This empathy helped deepen the nuance and accuracy of Moralign's key ethical protocols and standards, providing the impetus for Richard to be asked to lead the acting program. His current title was Vice President, Empathy and Interactivity, a title I gave him good-natured shit for whenever possible.

I tapped the glass, and Spat the robot looked in our direction.

"Dad," said Richard, slapping my hand. "Don't do that. This isn't a zoo. We're not supposed to distract him."

"Sorry." I fought the impulse to wave, even though I knew Spat couldn't see us anyway. "So what are we testing the little fella for today?"

Richard looked down and tapped the screen of his tablet. "We're going to introduce some common stimuli a robot kitchen assistant would deal with on a regular basis."

I heard the sound of a gas stove clicking on from the speakers above our heads. Spat reached for a pan and put it on top of the flames. "What's he going to cook?"

"A Thai chicken stir-fry," said Richard. "Spat's designed by Cuisinart since they've perfected a number of dietary algorithms based on people's budgets and food preferences. They hired us to get the human values programming in place so this latest model can be shipped to people's homes by Christmas."

"You think you can get him up and running that fast?" I asked. "It's already June."

"We should. They created Spat's operating system to be IRL compliant, so that speeds the process nicely."

"IRL?" I asked, confused. "In Real Life?"

"No," said Richard. "It stands for Inverse Reinforcement Learning.[2] It's a process created by Stuart Russell at UC Berkeley. Instead of creating a set of ethics like Asimov's laws of robotics—"

"Which were intentionally fictional," I interrupted.

"Which were intentionally fictional, yes, Dad, thanks," Richard agreed, nodding. "Instead of creating robotic rules based on written human values, the logic is that robots glean our values by observing them in practice. It's not semantics. Any values we wrote in English would have to be translated into programming code the robots would understand anyway. So reverse engineering makes a lot more sense."

I watched Spat as he sliced an onion, his movements quick and fluid like a trained chef's. "Still sounds hard."

"It is hard," said Richard. "But we base our algorithms and testing on a concept called degeneracy, which is the existence of a large set of reward functions for which an observed action or policy is optimal.[3] This aids our testing and heuristics to refine a robot's behavior until it overtly maps to a human value we can recognize."

I squinted. "Want to ungeekify that for me, Ricky?"

He frowned. "Don't call me Ricky, Dad, or I'll program your toaster to kill you while you sleep."

I laughed. "You can do that?" I smiled for a second, and then pictured a malevolent toaster in my bed like the horse head scene from *The Godfather*. "Seriously, can you do that?"

"*Anyway*, the point is we reward the robots for behaviors that reflect any human values we're trying to implement into their programming. That way we can reverse engineer preexisting code written in the manufacturer's language to dovetail with our patented ethical protocols."

"Well, that's freaking smart," I said.

"Yeah, it's pretty cool. We've also been able to come up with a lot of products on our own, based on the sociologists and behavioral economists we have on staff. A lot of times the ethical results that don't go as planned are as valuable as the ones that do." He pointed toward a theatrical set about thirty feet behind us, where a young married couple curled up on a couch was watching a movie. Like all of Moralign's environments, the small home was situated behind glass for optimized study. "Like with the Carsons. Last week they got in a huge fight after Shelia used the bathroom and Jack teased her."

"Why did he tease her?"

"Because she farted," Richard said, still keeping his eyes on Spat. "Jack was going to brush his teeth so he was outside the door. He made a joke and she got really pissed."

I shrugged. "Understandable."

Richard nodded. "Sure. Well, one of our sociologists remembered an old episode of *Freakonomics*[4] about the lack of sound in public toilets, and how in Japan they created something called the Sound Princess to mask the noise of people going to the bathroom. Apparently Japanese women kept flushing public toilets to avoid embarrassment and wasted a ton of water. So now we use a similar technology for Jack and Shelia. We're still tweaking the sensors. The AI algorithm was initially trained to be rewarded for turning on music when someone used a toilet. But that didn't work out a few nights ago when an AC/DC song turned on in the middle of the night when Shelia went to take a pee."

"Yeah, I'm thinking AC/DC is more of a mid-party kind of raucous pooping song."

"Yeah, we switched it to classical. Turns out Rachmaninoff works really great. Starts off low when you sit down and amplifies as needed in real time. We set up meetings with Spotify and Pandora next week to discuss potential updates to their ethical software programming. We think they should be able to tweak their algorithms fairly easily based on input from people's smart homes and toilets to include music in people's queues based on their dietary and bathroom habits, plus cultural preferences."

"Their *poo* cues?" I smiled. "You guys creating a new genre? Gassical music?"

Richard shook his head, sighing. "Seriously, Dad, I can have you killed pretty easily."

A mewing sound came from the speakers above our heads and we both turned as a fake cat entered the room near Spat.

"Thing looks like a furry Roomba," I said.

"That's what it is," said Richard. "Has a pretty basic algorithm based on a cat's movements. It doesn't need to look realistic for Spat. We just have to get him used to the presence of pets since so many seniors have them."

I nodded as we watched. Spat had finished chopping vegetables and I could smell the onions through a series of vents in the glass. He put a cover over the steaming vegetables and made his way to the refrigerator. The catbot got in his way, and Spat moved around him gingerly but with purpose.

I pointed toward Spat's body. "No legs?"

"No, this model is built to stay in the kitchen and the main floor of a home. His arms extend to reach high cabinets and he can move over carpets and floors. But he can't climb stairs. Makes him a lot more affordable."

We continued watching; the catbot's mewing growing louder. Spat opened the fridge and began pushing items out of the way, apparently looking for an ingredient.

"What's he looking for?" I asked.

"Chicken," said Richard. "This is one of the first tests for this new model. We want to see what the robot will do when it's confronted with data it wasn't expecting. In this case, when it chose its menu it scanned the smart fridge and saw the bar code of some chicken breasts we had in the freezer. So it chose the curry recipe based on that information. But one of my colleagues just removed the chicken a few minutes ago so now Spat has to update his algorithms in real time to satisfy his programming objective. In fact, my colleague removed all meat and tofu from the fridge and freezer, so the challenge is quite significant."

"What does that have to do with ethics?"

"Not sure yet." Richard looked at me. "It's more about taking an action that could reflect a value of some kind. But something always happens that gives us an insight in that direction."

"Cool." I noticed the catbot bumping against Spat's leg. "What about Kittybot? Is he trying to get Spat irritated? That could have moral implications."

"Robots don't get irritated, Dad. Just disoriented. But yes, we're trying to see how Spat will react with multiple standard scenarios."

The catbot extended a fake paw and began dragging it across the base of Spat's frame. In response, Spat closed the freezer door and moved toward a nearby cabinet, where he retrieved a can of cat food. He moved toward a kitchen drawer and pulled out a can opener, deftly holding it in his viselike claw. In three rapid twists, he opened the can, dropping the lid in the trash. Moving to get a bowl for the cat food, Spat held the can at his eye level for a long moment.

"Shit," said Richard, tapping his tablet quickly.

"What? He's reading the ingredients. Why, so he can make sure it's cat food and not poison or whatever?" I asked.

"Sure, but that's a simple check with bar codes or Beacon technology. The bigger test is we made sure this cat food is made largely from chicken. We want to see if Spat knows not to use it in his curry recipe since we took his other chicken away."

"Oh," I said. "Yeah, that would be less than savory. Not a fan of curry kibble."

We watched as Spat stayed motionless for another moment before reaching to get a bowl. He grabbed a spoon from a drawer and scooped out the cat food, placing the bowl next to the mewing Roomba. The robot hovered over the bowl like a real cat, staying in place while Spat moved back to the stove. By this point, the cooking vegetables smelled very fragrant through the vents, and the glass lid on the pan was completely clouded over with steam. My stomach made a small churning sound.

"No chicken, right?" I asked Richard.

He was chewing on his thumbnail, scrutinizing Spat. "No chicken," he replied, not looking at me. "Right now, Spat is reaching out to neighbor cooking bots to see if they have any chicken, as well as figuring out delivery times from FreshDirect self-driving cars or drones. But we timed this perfectly as a challenge scenario since this type of thing could happen in people's homes."

We kept watching. While Spat didn't actually move, I felt like I could see him grow tenser as the seconds clicked by. I knew he was a machine, but I also felt for the guy. Designed to be a chef, he risked screwing up a good curry and angering his owner.

A timer chimed in the kitchen, indicating that the chicken for the recipe should be placed on a skillet Spat had already pre-warmed and oiled. Quickly rotating 180 degrees, Spat bent at the waist and scooped up the mewing catbot, interrupting his eating. In one fluid motion, Spat placed the Roomba-cat on the red-hot pan, eliciting a horrified shriek from the fake animal in the process. Smoke poured from the stove, setting off an alarm and emergency sprinklers. A red lightbulb snapped on above our heads and Spat stopped moving as the scenario was cut short.

"Damn it!" said Richard, punching the glass.

"Why did he do that?" I asked. "Curried Roomba sounds pretty nasty."

Richard rubbed at his temples. "Spat reads him as a real cat, Dad. It means he saw the cat as a source of meat to use in his recipe."

"Ah." I sniffed, the scent of smoke and vegetables permeating the room. "You never know. Probably tastes like chicken."

Updating a Declaration

It appears that humanity's great challenge for this century is to extend cooperative human values and institutions to autonomous technology for the greater good.[5]

—STEVE OMOHUNDRO, "AUTONOMOUS TECHNOLOGY AND THE GREATER HUMAN GOOD"

Sensors and data have the capacity to easily take control of our lives. In any situation, we'll soon be able to analyze the emotions, facial expressions, or data about someone we're interacting with in business and social situations. While there will be great knowledge in this new cultural landscape, there's also great opportunity for emotional paralysis. How do you talk to someone when every word and emotion is analyzed? How do you form relationships of trust when you're worried about every thought and response? If we don't evolve values outside the context of this type of surveillance, our tracked words and actions will form the basis for preferences leading to a homogenized humanity.

This is why values play such an essential role in the creation of artificial intelligence. If we can't identify the human values we hold dear, we can't program them into our machines. Values don't just provide perspective on a person's life—they provide specificity. What morals do *you* feel are absolute? What spiritual, mental, or emotional qualities drive *your* actions?

If we were to create humanity's manifesto, what would it be? How can we codify such a treatise on human values to capture it for other people's use?

One idea would be to model an artificial intelligence ethics protocol after something like the United Nations' Universal Declaration of Human Rights.[6] Adopted in 1948, the declaration came about after the experience of the Second World War, when the international community came together to try and guarantee universal rights of individuals. The process took almost two years, and contains articles that could certainly be emulated for AI values, such as, "Everyone has the right to life, liberty, and security of person,"[7] or, "No one shall be subjected to torture or to cruel, inhuman, or degrading treatment or punishment."[8]

Where these values become difficult to implement, however, is in their lack of specifics. For instance, how should we define *liberty* with regard to programming machines? Would that mean allowing freedom for an autonomous program to iterate on its own, outside the confines of existing laws? Or does it refer to a machine's human operators, who must retain "liberty" of control over machines? As another example, what's "degrading" for a machine? Porn? Insider trading schemes?

Hernán Santa Cruz, of Chile, a member of the drafting subcommittee, had a response to the adoption of the declaration that's quite fascinating to reflect on regarding artificial intelligence and ethics:

> *I perceived clearly that I was participating in a truly significant historic event in which a consensus had been reached as to the supreme value of the human person, a value that did not originate in the decision of a worldly power, but rather in the fact of existing—which gave rise to the inalienable right to live free from want and oppression and to fully develop one's personality.*[9]

Let's unpack this a bit.

1. **The supreme value of the human person.** According to Santa Cruz, this value originates "in the fact of existing." In the context of AI, this mirrors the questions of human consciousness.

Does an autonomous machine deserve rights typically afforded a human just because it exists? That would mean Google's self-driving cars or militarized AI robots should be given those rights today. Or would a machine have to pass the Turing test—meaning at least 30 percent of the humans giving the test felt it was a person—to gain its rights? Or perhaps the machine would need to be self-aware enough to know it exists to gain these rights, like the androids in Philip K. Dick's *Do Androids Dream of Electric Sheep?*, the basis for the cult classic film *Blade Runner*. Whatever the case, the notion of a "human person" may soon be outdated or considered bigotry. So defining specific human qualities for AI ethics today is of paramount importance.

2. **A value that did not originate in the decision of a worldly power.** Whereas the value of a human person may originate outside the context of any particular worldly power, autonomous machines are currently being created in multiple countries and jurisdictions around the world. This means any AI ethics standards would have to account for multicultural views about the nature of the machines' existence. As always, the element of money clouds these issues, as there's such an opportunity for profit in the field of robotics and AI in the coming years. This means we need to separate the notion of value as *profit* from values as *human/moralistic guidelines.*
3. **To fully develop one's personality.** This phrase from 1948 is helpful regarding its specificity about a universal value. If it's possible to provide an environment where people could live free from want and oppression, Santa Cruz felt it was an inalienable right for those people to fully develop their personalities. Applied to an algorithm, would this tenet, again, imply that programs should be left alone to develop autonomously, outside of regulation? Or, as I believe, will the potential presence of multiple or even universal algorithms in our lives prevent us from naturally developing our human personalities?

While my initial vision for this chapter was to create an artificial intelligence ethics manifesto, I've come to realize it's impossible to simply provide a list of ten rules for all AI programmers to follow to ensure robots and machines are imbued with human values. As you can see from this example, we need a new Universal Declaration of Human Rights created in the context of how we're going to live with autonomous machines. This is why the petition created by the Future of Life Institute[10] around AI is so encouraging regarding the future. And while any declaration along these lines wouldn't be focused on mandating morals per se, it would reflect the moralistic ideas of the people who created them.

Difficulties in the Design

In my interview with Kate Darling, research specialist at MIT Media Lab, we spoke a great deal about the issue of ethics in AI programming. Her legal expertise and background in social robotics make her particularly suited to understanding the difficulties involved in creating standards for a field as vast as AI. One of the greatest challenges she points out is how siloed academia can be regarding other disciplines within the same institution and the world at large:

> *This isn't the fault of anyone building the robots. The message [within academia] is, "We don't want to restrict innovation. Let people build this stuff, then people doing the social sciences should work out the regulation after it exists." After I worked at MIT for a while, I realized there are very simple design decisions that can be made early on that set standards for later on that are very hard to change. You want people building these robots to at least have privacy and security in the back of their minds. But you bring these things up to them, and they often say, "Oh, I should have thought about that." It's a general problem that the disciplines are too siloed off from each other and there's no cross-pollination.*

Providing the cross-pollination Darling mentions is a much simpler solution to tackle regarding AI ethics than creating a universal standard of some kind. Fortunately, these issues have taken on a greater prominence since revered figures such as Elon Musk and Stephen Hawking have expressed concerns[11] over AI that have gained attention in the mainstream press.

Organizations within the AI industry have also tackled ethical issues for years, and in January 2015, the Association for the Advancement of Artificial Intelligence (AAAI)[12] even held its first International Workshop on AI and Ethics,[13] in Austin, Texas. Titles for talks at the event directly addressed issues of ethics, like the one presented by Michael and Susan Leigh Anderson, "Toward Ensuring Ethical Behavior from Autonomous Systems: A Case-Supported Principle-Based Paradigm."[14] The Andersons' hypothesis focuses on gaining the consensus of groups of ethicists regarding scenarios in which autonomous systems are likely to be used. Similar to the idea of observing human behavior I wrote about for the fictional company Moralign, the Andersons' proposal makes a great deal of sense as they point out in the abstract for their paper on the subject that "we are more likely to agree on how machines ought to treat us than on how human beings ought to treat one another." Once these agreements are reached and codified, they could begin to lay the basis for principles leading to ethical standards or best practices.

A significant difficulty in creating any ethical standards for AI comes with the nature of something James Barrat, author of *Our Final Invention: Artificial Intelligence and the End of the Human Era*,[15] calls "the inscrutability paradox." The basic concept has to do with the difference between what Barrat refers to as "designed" and "evolved" systems. Designed systems feature transparent programming, in which humans write all the code to allow for easier testing and scrutiny regarding ethical concerns. Evolved systems are outfitted with genetic algorithms or hardware driven by neural networks. Even before AGI (or sentient AI) is reached, self-perpetuating algorithms are far from uncommon. Analyzing these programs to allow for human intervention invokes the inscrutability paradox when their self-driving behavior can no longer be accounted for. The AI involved, even if intended to be "friendly" in nature, cannot be broken down for ethical analysis because it has evolved beyond human intervention from when it was developed. As Barrat notes, "This means that instead of achieving a humanlike superintelligence, or ASI, evolved systems or subsystems will ensure an intelligence whose 'brain' is as difficult to grasp as ours: an alien. That alien brain will evolve and improve itself at computer, not biological, speeds."

This is no trivial matter. In simplistic terms, it means turning systems off that have programming directives to stay on may become impossible. Not because an evil spirit has overtaken an operating system, but because the program is maximizing its efficiency through a logic that programmers would not be able to ascertain. This is why something like Steve Omohundro's Safe-AI Scaffolding Strategy[16] is so important to implement regarding ethical concerns, as it allows for iterated testing in which safe human intervention would still be possible at every step of the process. Fortunately, a good portion of the research priorities outlined in the Future of Life Institute's document regarding beneficial intelligence focus on security and control issues. All of Section Three of the document also considers how these issues will affect us in the future, including areas around verification, validity, security, and control for these systems.

"Being cynical, for ethics within all branches of technology, most researchers think of themselves as ethical and think about people writing about ethics as either being obvious or pompous. So in a field like AI, where it's very difficult to build intelligent systems, the concern that your system might be too intelligent and pose a risk hasn't been too high on people's agenda." This is a quote from an interview I did with Stuart Russell, author of one of the most widely used textbooks on AI, *Artificial Intelligence: A Modern Approach*.[17] I quoted Russell before, in his comment in response to the Jaron Lanier interview I mentioned in a previous chapter, "Mythed Opportunities," in which as a reminder Russell said, "Instead of pure intelligence, we need to build intelligence that is *provably* aligned with human values."[18]

I am greatly encouraged to hear that this thought leader feels AI programming needs to be aligned with human values even with "unintelligent" AI systems. This means, as he notes, changing the goals of the field to incorporate the perspective that human values provide beyond the general notion of intelligence.

A recent event involving Facebook algorithms provides a great example of what I mean in regard to the importance of aligning human values with algorithms. In December 2014, Facebook introduced a feature called Year in Review, which allowed users to see their top pictures and posts from the past twelve months based on friends' clicks and likes. The algorithm creating the feature also took a person's pictures and posted them within a holiday frame featuring happily dancing cartoon characters. On Christmas Eve 2014, Eric Meyer, author and founder of web design conference An Event Apart, wrote a post for his blog called "Inadvertent Algorithmic Cruelty."[19] As it turns out, Facebook's Year in Review feature posted a picture of Meyer's daughter in his feed, not recognizing that she had passed away six months before. Meyer's response points out the need for ethical reasoning behind the systems already driving so much of our lives:

> *Algorithms are essentially thoughtless. They model certain decision flows, but once you run them, no more thought occurs. To call a person "thoughtless" is usually considered a slight, or an outright insult; and yet, we unleash so many literally thoughtless processes on our users, on our lives, on ourselves.*[20]

By definition algorithms are thoughtless and heartless. It doesn't make sense to try and define code by human values in this regard, since they're completely separate paradigms. This is why Stuart Russell's work regarding Inverse Reinforcement Learning (IRL) is so compelling. He was the inspiration for the scenario opening this chapter, based on our interview for this book. During our discussion, I mentioned the well-known example from AI ethics about an algorithm designed to make paper clips, as discussed in a previous chapter. While it seems harmless enough in and of itself, if the program were set to create paper clips at all costs, it might harness electrical power from nearby buildings or hoard other natural resources needed by humans to satisfy its primary directive. As compared to a simplistic, "AI machine gone rogue" scenario, the example is used to show people the importance programming plays in designing autonomous machines.

Russell, however, points out that goals for humans exist within the context of how we've already lived our lives up to the point we receive a new goal. "When I tell a human to make paper clips, that's not what I mean. I want you to make paper clips in the context of all the other goals you've ever been given and that everyone else takes for granted within the spectrum of morals and goals we all have." This is why Russell feels there should be companies that construct representations of human values, including this concept of people's backgrounds that would recognize the layers of ethics, laws, and morals we all naturally take for granted. That's where I got my idea for Moralign, as well as the punch line of the story about cooking the cat: "If there's nothing in the fridge," noted Russell in our interview for *Heartificial Intelligence*, "the last thing you want your robot to do is put the cat in the oven. That's cooking dinner—what's wrong with that?"

I think the notion of IRL that Russell is working on provides a solid methodology for creating a set of ethical standards for the AI industry. Like the work of the Andersons, mentioned previously, by observing how autonomous systems respond to and interact with humans, we can more easily determine how people should be treated than by simply philosophizing about future scenarios. It's also encouraging to note Russell does believe more experts are beginning to address ethical issues as AI is already having such a huge impact on society. However, the problems of silos and vested interests will still have to be dealt with to ensure human values are universally reflected: "AI is a technology where the people who develop it aren't necessarily the people who use it and the people who use it have the best interests of their shareholders or their secretary of defense at heart. Outcomes may not be what the human race would want if we all put our heads together."

Crowdsourcing Control

AJung Moon[21] is a Ph.D. candidate studying human-robot interaction and roboethics with the Mechanical Engineering Department at the University of British Columbia. She is also founder of the Open Roboethics initiative[22] (ORi), an organization that allows a multidisciplinary community to crowdsource people's opinions on ethical and moral issues surrounding emerging technology. Crowdsourcing and collaboration allow website visitors to suggest polls based on ethical questions the community can vote on, such as issues around autonomous vehicles or elderly-care bots. Her work and the crowdsourcing model provide a pragmatic and compelling way to examine issues of ethics in artificial intelligence.

I interviewed Moon about an experiment she conducted featuring a delivery robot and a set of scenarios involving how the robot would deal with humans while waiting for an elevator. In the video featuring the experiment,[23] viewers are presented with multiple scenarios mirroring the types of ethical decisions we currently make while waiting for an elevator in a crowded building. As Moon explains in the paper[24] describing the experiment, "The goal of this work is to provide an example of a process in which content from a collection of stakeholder discussions from an online platform can provide data that captures acceptable social and moral norms of the stakeholders. The collected data can then be analyzed and used in a manner suitable to be implemented onto robots to govern robot behaviors." In other words, Moon believes standards for AI and robots can be created by humans crowdsourcing their aggregate opinions around certain situations to form an ethical framework that can be adopted by designers.

Her work and the community's polls are fascinating in regard to the complexity and depth of human ethics we need to examine in light of AI technology. For instance, in the video series showing the large delivery robot next to a person in a wheelchair, the robot offers to take the next elevator, which is how most of us would likely react. But would the person in the wheelchair see this as a form of condescension? And how should the robot respond to the person in the wheelchair in a country where women are second-class citizens? Will manufacturers provide a "human base level" set of ethics that can then be iterated based on local culture in different countries? Taking part in ORi's polls is an excellent way to begin to understand and empathize with the complexity of decisions faced by AI manufacturers while also confronting the urgent need for individuals to become ethically self-aware. As Moon noted in our interview: "Looking at a very simple, daily-life decision scenario, we can come to a consensus so we can program a type of human and democratic decision making into the system. Our purpose of conducting these polls is to involve the general public in learning what kinds of things people value."

The Effect of the Ethics

Technology already exists that can measure facial expression as a proxy for emotion. Sensors in and outside our bodies will soon be able to enhance the capabilities of algorithms running Facebook or other services we use throughout the day. To some degree, we've all had experiences with machines or software like the one Meyer described regarding his deceased daughter. This is because we're still able to differentiate between machines or poorly designed algorithms and the humans in our lives. But this is a finite era. While we may recognize the glitches associated with newer technologies, we also tend to forget the multiple times we've responded to the voice of our GPS as if it were a real person or the reverence we direct toward our mobile devices. Our values regarding the ubiquity of technology in our lives have already fundamentally shifted. Now with artificial intelligence on the rise we get the unique opportunity to decide what parts of our humanity we feel are worth automating. Or not.

This process is about much more than standards or regulations. I'm not interested in creating a set of rules just for the sake of clarity or legal purposes. If we are truly at the end of the human era or at a point in our evolution where machines may gain a prominence in our lives like never before, now is a great time to illuminate our manifesto on humanity.

I interviewed Steve Omohundro for this book to discuss his ideas on ethics and AI. I closed our conversation with a final question I often like to ask interviewees, which is, "What's the question nobody asks you that you wish they would?" I do this because many times experts like Omohundro get asked similar questions based on their most popular theories and I always wonder what's on their minds they think journalists may have missed. Here's what he had to say: "I don't see a lot of people asking, 'What is human happiness?' or 'What is the model of a human society?' People often wonder if AI is going to kill them, but they don't think about the fact that if we had a clear vision of what we're going for with these bigger types of questions, then we could shape technology based on the vision of where humanity should go."

Here are the primary ideas from this chapter:

- **Human values should be central in the creation of artificial intelligence.** Ethics as an afterthought won't work in the widespread adoption of autonomous systems. "Evolved" AI programs that won't allow human intervention make ethical standards useless unless incorporated at the earliest stages of development. As Stuart Russell notes, these values-based directives should also be applied to non-intelligent systems to shift the goal of the AI industry from seeking generalized "intelligence" to outcomes that can be provably aligned with human values.
- **It's time to break the silos.** While it's common for silos to exist within academia, researchers, programmers, and the companies funding their work have an ethical responsibility to break down these barriers in regard to the production of AI. In the same way sociologists adhere to standards in creating surveys or other research to study volunteers, developers need to apply

similar criteria to the machines or algorithms they're creating that directly connect with human users.

- **AI needs to incorporate Values by Design.** Whether it's IRL or a different methodology, issues regarding values and ethics need to be a standard starting point for AI developers in academia and corporate sectors around the world.

ten | Mind the GAP (Gratitude, Altruism, and Purpose)

Winter 2022

I dealt with a lot of depression and stress regarding Melanie and her chip. But with all my research into happiness and well-being, I couldn't just wallow in negativity. I made a joke along these lines to my friends who also worked in areas of well-being, saying that would be "off-brand." We were allowed to be human and have bad moods, but we couldn't be misanthropic. So I knew I had to take steps to deal with my issues around the chip, especially so I could focus on helping Melanie and Richard versus focusing on myself.

So after Melanie got her implant, Barbara and I did a lot more research into deep brain stimulation and the work of people such as neurosurgeon Andres Lozano[1] to learn about the latest treatments for Parkinson's, as well as the stress and depression it could bring. While most of my work was focused on hacking happiness, people such as Lozano and other experts were exploring how to hack the brain with electricity.[2] The basic concept was that small jolts of electricity targeted at specific parts of a person's brain could help treat Parkinson's or epilepsy. Sometimes, as in Melanie's case, a patient needed a chip directly in his or her brain for greatest results. However, a relatively noninvasive technique known as transcranial direct-current stimulation, or tDCS, was gaining in popularity among the general public, as equipment to administer the process could be purchased easily and used at home.

In my case, I purchased a Foc.us headset[3] I'd read about online that claimed to increase performance for video gaming and working out and potentially would decrease depression.[4] The electrodes looked like old-school headphones, with a plastic coating versus softer material to place over your ears. The logic of the electrodes was that the electrical impulses going to specific places, such as your cerebral cortex, enhanced the natural synapses within your brain.

I'd read about a number of other people's experiences using the headset, and nobody had suffered any long-term negative effects. Some people got headaches or mild nausea using it, but that can happen during normal exercise or while watching *TMZ*. My experience was quite enjoyable using Foc.us. I played one of Richard's PC games before and after using the headset, and actually did better while using it. Richard teased me for looking like a total dork the whole time, but I felt a mild physical sense of euphoria and increased confidence with it on.

The trouble was, I kept thinking about something other users of the headset had mentioned—the placebo effect. Similar to the nature of bias in prescription drug testing, the placebo effect meant my belief that the headset was working may have brought about its apparent success. The headset could simply be a fancy gadget helping me invoke a sense of self-esteem. I'm pretty sure it worked on its own, but there's no way to avoid the rabbit hole of being biased about testing something. It's like being told to sit in a room and not think about an elephant.

Besides, as Melanie pointed out, she had her chip out of necessity. When her doctor pointed out benefits she might experience in her situation such as the ones Foc.us offered, she would just nod. "I just pretend the chip is part of my brain, Daddy," is how she explained it to me. "I can't walk around all day wondering if my thoughts are special or whatever. It would drive me crazy."

She was amazingly mature for her age.

She was also the reason my well-being improved after her surgery. I had never been so grateful for anything in my life than to see her healthy after her chip was implanted. I still worried about her getting hacked, or playing with magnets, or any number of other things any sci-fi paranoid could conjure up. But she was here. She was with me. And from all outward appearances, she was like any other girl her age.

In my work studying positive psychology, I've learned about the power of gratitude. It isn't simply fluff—expressing thanks for the things you have versus lamenting what you don't has been scientifically proven to increase your well-being.[5] However, the benefits of gratitude come *after* you take action to be thankful for what you have. This is where people mistakenly think gratitude is like happiness—that it can only be real when it organically arrives as part of your day. While the mood that accompanies gratitude is a bonus, the science shows your well-being improves *after* practicing thankfulness in the same way benefits from the gym accrue only after you actually show up and sweat for a while.

But the point is not to get caught up in semantics or academia. The goal is to make a list of the people and benefits you have beyond money or other people's perceptions and be fully aware how blessed you are to have them. This brings humility and a deep sense of thankfulness. I've discovered a number of apps that are focused on gratitude, such as Gratitude 365,[6] that help you keep a journal of things you're thankful for throughout the day. It's a great tool, and got me used to aggregating my personal insights in a way I could benefit from more than my posts on Facebook, which I forgot about within a few days after writing them. But while this quantification provided details about the people and things I valued, I still had to get used to putting down my phone and purposefully practicing gratitude.

For me, gratitude is a tool to improve myself so I can be better for others. I find that if I express gratitude via Facebook or Twitter, I may get caught up in showing off that I'm being grateful or am trying to be altruistic toward whomever I'm talking about in my posts. It gets a little complex at times, but fundamentally I believe gratitude is a personal endeavor in which you gain the most benefit by choosing to savor the elements of your life for which you are grateful.

I don't want to give a false impression, however. I didn't come to understand the benefits of gratitude until I forced myself to practice it. I read the science and thought it made sense, but still had to make it work in my own life to know it wasn't just feel-good crunchy philosophy. So I hacked my Foc.us headset and used it to measure dopamine[7], oxytocin, and serotonin[8] levels in my brain before, during, and after my gratitude sessions. These are often referred to as the "happiness hormones," and they would increase only if I got to a meditative place of reflection and gratitude. In other words, I couldn't cheat and just look at a list of names I'd written down to say I'd practiced thankfulness for the day. Geeky accountability can be effective.

I modeled my sessions after some mindfulness practices that involve a form of meditation in which you deeply take in your surroundings and concentrate on your breathing. I'd done a lot of these types of exercises as an actor, and they basically help you relax so you can savor the things you've identified being grateful for. It's amazing how much discipline it takes to avoid mental distraction. It's ironic that our technological age has given us so much access to information, but if we don't gain focus it slips away before it becomes wisdom.

I'm kind of an all-or-nothing sort of guy when it comes to things like exercising or dieting. I don't necessarily recommend this mind-set, but I don't have the willpower to work out just a little each day hoping for positive results. Back in 2014, I lost thirty pounds in four months by working out a few hours every day because I know how I roll, and I have to go overboard to jump-start a habit. In the case of gratitude, I used an app that wouldn't allow me access to the Internet or files on my computer until my hormones spiked (meaning I had gotten my grateful on). I also hardwired the app into my smart house and car so I couldn't use the microwave or drive if I tried to shortcut my gratitude sessions. Thankfully, by the fourth or fifth day, I was so enjoying the benefits of gratitude that I didn't need these tricks to savor what I had in life. So now I don't use the headset any longer. If I get too busy, or have to travel, or whatever, I stare at a picture of the kids and Barbara, to which I taped a sticky note that reads, "gratitude." Simple.

Once I got a handle on gratitude, I began practicing a form of ethical altruism in my life. There's a great deal of scientific evidence[9] showing the benefits of altruism, including things like increased happiness, decreased stress, and an increased sense of self-esteem. When I say "ethical" altruism I mean I try to base my actions on the values I hold dear as well as the skills I possess. I realize this causes consternation for some, who believe altruism is degraded if it's too subjective or isn't motivated by selfless compassion. I get their logic, but think it's counterproductive for two reasons: First, I'm always subjective in the sense that I'm always me; and second, I love how helping others makes me feel. Besides, if you actually help someone, I think it's largely an academic exercise to figure out if you were being "truly selfless" or not. This implies rules of some kind—you should only give something to someone anonymously, you should only give something to someone where others won't see you, etc. You definitely should be tactful and give with grace, versus being showy for public acclaim. But I'd rather not add constrictions to people desiring to increase their altruism. Give with your heart, give often, give like you'd want to receive. Help where it's needed now versus talk it to death and risk someone suffering needlessly in your absence.

To get started on my road to proactive altruism, I discovered an app called Heroes,[10] which was featured in a contest to create happiness apps from an organization called ChallengePost (now Devpost).[11] Designed to be used to foster community outreach and provide disaster aid, the app lets people upload needs or emergencies to gain assistance from local friends or neighbors. It leverages mesh networks from people's Bluetooth and Wi-Fi, which also means it doesn't need an Internet connection to work. In my case, I started using it to help my neighbors, since you can post simple things like, "I need a ladder" as well as, "I need a ride to the hospital." It's essentially a manifestation of the sharing economy plus a more overt way for people to express what they need and others to help them out based on their skills. For me, the app served as a conduit for altruism and, as the name implies, helped me feel like a hero while getting me out of the house.

In 2020, my altruism brought benefits I hadn't expected when President Elizabeth Warren introduced a universal basic income bill that, surprisingly, passed through Congress. Also called a basic income guarantee (BIG), this type of income plan operates on a fairly simple idea: Every citizen gets a base amount of money that allows them to survive, whether or not they choose to work. While it may seem shocking that conservatives would go for such a bill, Warren was able to demonstrate via data from Canada[12] and Switzerland[13] how creating a wage floor is an effective way to fight poverty and reduce government spending.[14] I first learned about BIG via a video from the *Washington Post*'s *Wonkblog*[15] called "Is giving everyone a check a good idea?"[16]

Years ago, in the pre-Internet days, when I used to be an actor, I received unemployment benefits once or twice after going in person to register. It was embarrassing and dehumanizing seemingly by design. Rather than feeling my country was supporting me to seek more work while avoiding homelessness, I felt the almost visceral presence of Puritan forefathers chastising me for my laziness and lack of moral rectitude. It sucked. I half-expected a reanimated Horatio Alger to kick my ass in the alley behind the office after I'd filled out my forms. Basic income guarantee plans, while designed to provide a form of welfare, differ from "the dole" in the sense that money is provided to all citizens. This is meant to help eradicate the stigma of receiving "handouts" from the government.

Ultimately, the goal of a BIG is to help people, regardless of current job status, race, or gender, get by on their own and keep them from straining public resources. This may seem like a crazy idea and it's easy to be concerned about people who will opt not to work out of laziness or lack of a work ethic. But automation by machines has put a great deal of the population out of work, which means America's GDP has been suffering the past number of years. People can't be good consumers without having money to purchase things. Go figure.

The idea seems less crazy when you think about it in terms of something called a negative income tax (NIT).[17] As Philip Harvey, of Rutgers University, notes in his paper "The Relative Cost of a Universal Basic

Income and a Negative Income Tax," "An NIT is a system of refundable tax credits that guarantees eligible tax filers a certain minimum income. Tax filers with no income from other sources receive the full NIT benefit in cash, thereby providing them a basic income guarantee (BIG)."[18] This comes with its own consequences, such as a potential for countries with BIGs to try and keep immigrants from gaining citizen status. But in Warren's case, her focus is on eliminating poverty while spurring the economy for all citizens rather than providing tax breaks or exemptions largely to highly paid individuals or corporations.

Warren also included a provision in the bill that allows individuals to increase the size of their guaranteed incomes based on contributions to their communities or society at large. This was an effort to provide more direct funding than was often available to individuals outside the non-profit world. A few states also opted to create programs that would increase any parent's income whose children participated in approved community service programs. This was based on the ideas of Kathleen Kennedy Townsend, daughter of Robert Kennedy, who had created mandatory community service programs[19] in Maryland before the turn of the century. While there was concern this might cause deadbeat parents to force their kids to work, programs were carefully monitored to avoid any fleecing along these lines. Students also recognized that their efforts provided a form of job training, or they could opt to have monetary equivalents based on work output put in a savings account to be used for college or added to their BIG in years to come.

For my part, I opted in to a program called Community Altruism and Respect for Elders (CARE) after Warren's bill passed. Using my writing and storytelling skills, I now go to seniors' homes or living facilities and write down their life stories for posterity. I video record them at the same time to provide archival footage for their families or for future knowledge bases. It's a service I used to do for corporate clients, but seniors and their families wouldn't have been able to afford me without this type of program assistance.

Sometimes the seniors will have a robot companion such as Pepper or

Jibo in the room when I visit. The robots can record video and interface with human staff in a hospital, and I'm getting used to them. A lot of times, however, the seniors will ask me to turn off their robot companions when I arrive after they've gotten used to me. It's quirky, but in those situations I feel like a sort of ex-boyfriend or -girlfriend in front of the new spouse. I know the robots don't actually feel slighted, but turning them off definitely increases the human intimacy between me and my clients.

It's easy for me to rant or get political about automation and how BIG is a great solution. It's just as easy for other people to rail against what they see as handouts. All I know is that since the program has started, a lot of people are a lot less terrified of not being able to take care of their families. As a history fan, I often wonder if I'm experiencing the same joy workers must have felt during the New Deal with Franklin Delano Roosevelt. While I get the idea of a Puritan work ethic, I'm not only excited to be working with seniors because I'm getting paid or have a job—I'm flourishing because my work is based on my skills, so I have a sense of purpose beyond providing for my family. I can't control the proliferation of AI and automation, but I can savor my time with clients recording the stories that comprise their lives. And here's a little secret: It helps me as much as it helps them.

Happinomics

> *The future belongs to the altruists. We are born with the predispositions necessary to maintain ourselves in the world. But while we are familiar with the rationally justified pursuit of our own advantage, we are still uncertain about the impulses that lead us to seek our own happiness in the happiness of others.*[20]
>
> —STEFAN KLEIN, SURVIVAL OF THE NICEST: HOW ALTRUISM MADE US HUMAN AND WHY IT PAYS TO GET ALONG

Our well-being and artificial intelligence are inexorably linked. While it may take decades for machines to achieve sentience, algorithms and data are already omnipresent in our lives. The threats of automation and job replacement are very real, so we need to seriously consider how we'll be happy as humans if we don't work.

My solution is something I call happinomics, a combination of tenets from positive psychology and measurable, actionable metrics. I modeled the positive psychology portion of this idea on my fondness for all things British, utilizing an acronym modeled after the Tube (what they call the subway in the UK): Mind the GAP, in which GAP stands for "gratitude, altruism, and purpose." I break down purpose into two parts: values and flow. Here's how I explain each of these elements:

- **Gratitude:** As I alluded to in the opening narrative, gratitude is a tool to help center yourself. It encompasses ideas of mindfulness and savoring, and acts as a sort of emotional training program to increase your well-being.
- **Altruism:** Sometimes referred to as compassion, altruism provides an opportunity to extend one's emotional training/well-being to others.
- **Purpose:**
    - ✧ **Values.** As described in the values chapter, I believe the specific ethics, morals, and values that drive a person's life provide the specificity to define what makes him or her human. These ideals implanted into our machines will ensure we continue to benefit from the attributes that separate us from machines. They also provide a way to identify what's important to people on an individual, community, and country basis.
    - ✧ **Flow.** As I described in *Hacking H(app)iness*, Mihaly Csikszentmihalyi's *Flow: The Psychology of Optimal Experience*[21] is one of the seminal books in the positive psychology lexicon, and has been a national best seller since it was first published in 1990. While flow isn't always pleasant—athletes may be in a physical state of agony while achieving optimal

> experience, for instance—it represents our state of being when we're doing something we feel we were built to do. It also often contains an almost insurmountable challenge that brings deep satisfaction, since in mastering a skill you gain a profound sense of meaning and achievement. As Csikszentmihalyi notes in *Flow*, "The best moments usually occur when a person's body or mind is stretched to its limits in a voluntary effort to accomplish something difficult and worthwhile."[22]

I've created a few simple exercises you can try to experience these attributes in action. They're designed to be done with a partner:

GRATITUDE

- Person A asks Person B: "What are you grateful for?"
- Person B responds in two or three sentences; e.g., "I'm grateful for my family because . . ."
- Person A repeats why Person B is grateful: "You're grateful for your family because . . ."
- Now repeat the process, switching Person A and Person B.

Being grateful helps you be mindful and appreciate what you have versus what you don't. It can also be very powerful to hear someone else remind you of what you're grateful for, even if he or she is simply repeating your words. It gives you an empathetic face to associate with the specific areas you're thankful for in your life.

ALTRUISM

- Person A asks Person B: "What's one idea you've come up with at work or at home you feel is worth sharing and why?"
- Person B responds in three or four sentences.

- Person A compliments Person B with specific ways he or she thinks this idea is helpful.
- Now repeat the process, switching Person A and Person B.

"ROA" (return on altruism) is twofold: You increase your self-esteem while helping others. I included the quote from Stefan Klein at the beginning of this section, as I feel it's imperative to move toward a robotic future knowing how to "seek our own happiness in the happiness of others." While social robotics and AI certainly improve the quality of our lives, altruism and compassion offer us free ways to achieve well-being today and into the future. These actions cost time and a shift in cultural attitudes, but they offer alternatives to consumption of goods or pharmaceuticals as a way to offset depression. In this sense, as with gratitude, personal interventions regarding our own and others' emotional well-being could have potent economic ramifications at scale. And while companion robots can mirror or replicate compassion, we humans cannot gain the increase in intrinsic well-being unless we take these actions for ourselves.

VALUES

- Using the values survey offered in Chapter Eight, consider how your interests and ethics could align with the needs of your community. Do you genuinely feel helping others according to your values could increase your well-being? Are you willing to try and test these ideas?

FLOW

- Person A asks Person B, "When is the last time you lost yourself in work or an activity?" Losing yourself is called "flow"—doing an activity you were built for, as mentioned.
- Person B responds in three or four sentences.

- Person A asks, "How would spending more time doing that work or activity improve your life?"
- Now repeat the process, switching Person A and Person B.

Doing the work you were built to do increases your intrinsic well-being. While we've all had to do money jobs to pay the bills, flow can be obtained by doing things such as learning how to play a musical instrument or getting in better physical shape. The point is to identify the activities you feel you were born to do and identify how you can increase them in your life to amplify your well-being.

I picked the term *GAP* because it also refers to the idea of a gap analysis in the business world. *Wikipedia* defines a gap analysis as "the comparison of actual performance with potential or desired performance. If an organization does not make the best use of current resources, or forgoes investment in capital or technology, it may produce or perform below its potential."[23]

Currently, most of us base our sense of self-worth on our wealth, and GDP is designed to measure a country's happiness via financial metrics alone. These measures by themselves don't encompass all of who we are. They are not sound proxies for well-being, hence the GAP, which is to be filled by measurable actions focused on gratitude, altruism, and purpose.

Is this a simplistic idea? Yup. Is it idealistic? Yupper.

But it's also testable. The Internet of Things now includes the notion of measuring people via their health data, emotions, and actions. As I showed at length in *Hacking H(app)iness*, the quantified self and wearable industries have recently exploded in popularity. Whether individuals take the time to track themselves or automatically have data measured by things like Apple's Health app, it's shortsighted to think these new paradigms of data that deeply reflect emotional and physical well-being won't influence business, government, and the culture at large. While today my talking about tracking happiness or measuring gratitude may sound quaint or crunchy, there's simply too much data available about individuals based on these attributes to ignore. Happiness based on increased consumption is not only ineffective, it's out of style. This is why new metrics such as gross national happiness and the genuine progress indicator provide better methods of tracking citizen well-being for the future, as we'll discuss in the next chapter.

For now, let's keep things simple. If you tried any of the exercises I included in the previous pages, did they improve your well-being at all? If so, great. I've included links at the end of the book to find out more about all of these areas, including things such as the Science of Happiness[24] course offered by the Greater Good Science Center,[25] and the online community Happify,[26] which provides multiple positive psychology tactics to measure and improve your well-being over time. While you need to explore these resources on your own to identify the best tools that work for you—similar to an exercise regimen—I can guarantee any of them will improve your well-being more than:

- Getting replaced at work by a robot.
- Arguing about when you'll get replaced at work by a robot.
- Wondering what job you'll have after you get replaced at work by a robot.

I also believe people utilizing the skills that bring them flow based on their values will help transform economies at scale. Big Data could be used to identify where communities have needs and which individuals could serve them. Like the Heroes app, people could gain a sense of purpose and increased self-esteem without needing money. This is the type of happinomics I hope will someday soon become a reality alongside the inevitable rise of AI.

The BIG Idea

I brought up the idea of a basic income guarantee in my opening narrative, as we can't consider issues of well-being in the near future without coming up with potential pragmatic solutions regarding the onset of machine automation. As previously discussed, Moore's law and the left-hand-turn problem with Google's self-driving car both demonstrate that the human attributes we believe machines or AI can't replace are diminishing rapidly. While it may be true machines are currently better at analytical tasks than ones requiring empathy, the field of social robotics focuses almost entirely on how humanlike interactions can provide a sense of companionship and well-being. So it's safe to assume machines will at least be able to *pretend* they're empathetic and emotional, which is more than a lot of us humans can do.

In terms of pragmatic solutions regarding automation and AI in our near future, Stan Neilson, author of the book *Robot Nation: Surviving the Greatest Socio-economic Upheaval of All Time,*[27] provides a no-nonsense, solutions-driven treatise on how to deal with what he sees (as I do) as the inevitability of machine automation. Here's the opening paragraph of his book:

> *Whether we like it or not, we humans are destined to become obsolete. This will happen as soon as intelligent robots exceed our capabilities. This will certainly occur within the next two centuries, but it will probably happen much sooner. After we become obsolete, I see only three possible futures for humanity. We might somehow rule a race of subservient robots that do our bidding despite our physical and mental inferiority to them. We might be tolerated and controlled by a race of robots that is indifferent to our existence or finds as useful as slaves, pets, or objects of study. Or we might be exterminated by a race of robots who find us too dangerous and too counterproductive to their ends to be allowed to exist. In short, we might be worshipped, enslaved, or exterminated.*[28]

Wa-bam! How's that for an opener? Neilson is a former AI engineer with a background in developing military and commercial applications. If someone who's an expert in the field sees only these three possibilities for our future, I'm inclined to listen to what he has to say. The book is meaty and fascinating, and I mention it in this chapter because it offers a sense of how to straddle the ethical dilemmas associated with AI and the need for humans to have a sense of purpose to achieve well-being. Neilson identifies this need as a fundamental ethical objective that's a form of modified utilitarianism. Utilitarianism is a theory in normative ethics focused on maximizing utility, or as Jeremy Bentham—one of the greatest proponents of utilitarianism—said in his famous treatise, "A Fragment on Government," "It is the greatest happiness of the greatest number that is the measure of right and wrong."[29] This is sometimes called the "greatest happiness" principle; Neilson offers a form of what he calls this "modified Benthamism" as a way to create ethical standards for AI.

As I stated in the Introduction to the book, I'm not an ethicist, but I am fascinated by the discipline's focus and application of morals, philosophy, and action. According to *Wikipedia*, normative ethics is "the branch of philosophical ethics that investigates the set of questions that arise when considering how one ought to act, morally speaking. . . . [Whereas] descriptive ethics would be concerned with determining what proportion of people believe that killing is always wrong . . . normative ethics is concerned with whether it is correct to hold such a belief."[30] In this regard, utilitarianism makes a great deal of sense when considering how to create standards for AI that can reflect our values. Its focus is on the consequences of actions, which can be more easily studied than a person's intentions or morals. This is also why I believe tracking our values is so critical, as it allows us to see how we actually live out our claimed beliefs versus our stated best intentions.

Neilson's analysis of how his modified Benthamism could play out in pragmatic terms is fairly exhaustive, and he points out the theory's inherent moral challenges. For instance, utilitarianism would argue that "maximum happiness" could be achieved when a child's torture provided a "cure for all the world's ills."[31] This idea is morally reprehensible, but it's a highly relevant example of what ethical standards committees are facing today regarding militarized AI. From a pragmatic standpoint, is it better to target and kill fewer individuals or accept the loss of civilian lives as "collateral damage" in war? Answer that question and you're beginning to see the complexities of creating ethical standards for AI.

But it's a starting point. Combining ideas from Stuart Russell and his reverse-engineering mind-set of IRL, we could start to pioneer a sense of human purpose in the midst of AI and automation taking place today. But while minding the GAP may empirically show that people can improve their well-being via positive psychology, these actions won't be as effective if people don't have jobs and the sense of purpose they provide.

In my interview for this book with Martin Ford, author of *The Lights in the Tunnel: Automation, Accelerating Technology and the Economy of the Future*,[32] he advocated for the idea of a guaranteed income as I described in my opening narrative. He felt it could provide a future where humans could attain a sense of purpose beyond traditional jobs. Ford pointed out in our conversation that the BIG idea is not a new one, but was championed by luminaries such as Milton Friedman and Friedrich Hayek, and that Richard Nixon actually proposed a form of this type of guaranteed income in his Family Assistance Plan of 1969, providing a stipend for all American families with children.[33] Here's how Ford described his thoughts to me on the subject:

> *I think we have to move forward toward a guaranteed income. This would need to include an incentive for people to pursue education for eligibility, so their incentive to learn doesn't go away. You could also have an incentive to work in the community or do something positive for the environment. These would replicate aspects of a traditional job when income is decoupled with work, which is where I think we're heading in the future. By offering some sort of sense of purpose to people, they'll attain a sense of fulfillment even though they won't have traditional jobs.*

This idea makes empirical sense, as positive psychology shows people's intrinsic well-being increases when they use their skills independent of whether or how much they're paid. As an example, Ford mentioned how much labor and time people put into *Wikipedia* as contributors even though they don't receive money for their efforts. Their incentives to contribute to the site aren't focused on material gains, but on the sense of purpose they receive in the process.

A huge consideration for adopting a BIG has to do with motivation. Why would people continue to work if they got paid for doing nothing? For one thing, the majority of these plans recommend providing only a sustenance-level amount of money to individuals. The idea is that most people would continue to work but have a "safety floor" of money, the equivalent of a savings account today. This financial protection, in the wake of widespread automation, would provide a pragmatic solution toward helping people stay afloat while they sought new jobs or activities to increase their well-being. I'm sure there would be "slackers" or individuals who chose to neither work nor help others with these BIG solutions. But that inevitability doesn't provide us the luxury of ignoring the vast numbers of humans who *will* want to work and find purpose in a future where jobs aren't available.

Marshall Brain,[34] host of National Geographic's *Factory Floor*, provides a compelling narrative about the future of automation in his series of essays called *Robotic Nation*.[35] In Part Three of the series, "Robotic Freedom,"[36] he outlines a number of the "traditional" solutions currently posed by experts on how to deal with widespread automation. These include ideas such as banning robots from the workplace, reducing the average workweek, and taxing robotic labor, all of which he discounts as being unrealistic.

Instead, he offers the idea of "supersized capitalism," similar in nature to a BIG solution, in which all citizens would receive twenty-five thousand dollars a year to spend in the economy. This would keep the economy strong, spur innovation and entrepreneurship, and ensure people could survive in the wake of automation. He then provides sixteen different ways this money could be obtained, including things such as a national mutual fund, "sin" taxes (in which money from products such as cigarettes and alcohol would go toward the twenty-five thousand),

and "extreme income" taxes to address the country's vast concentration of wealth among corporations and the wealthy. Brain's multipart essay "Manna"[37] provides a highly compelling fictional narrative regarding automation, in which technology driven by profit motive alone inevitably leads to human misery.

Whether these solutions seem tenable or not, they provide implementable ideas that represent situations we're going to have to deal with regarding AI, automation, and how we'll achieve well-being and happiness in the future.

Purposeful Pragmatism

It's easy for me to be pessimistic about our future regarding AI. But I can't advocate minding the GAP to you and personally wallow in fear or negativity. These emerging technologies are creating wondrous advancements, and we should embrace them, just not in isolation. I will be thrilled to be proved wrong regarding automation replacing all our jobs, especially considering the future of work for my kids. But planning for those potentialities in the future is still an imperative, which includes embracing the ideas of positive psychology. That way we can train ourselves how to find purpose without the need for a traditional job, and be grateful in the process.

Here are the primary ideas from this chapter:

- **It's time to mind the GAP.** Empirical science continues to demonstrate how our well-being can be improved via actions such as expressing gratitude, being altruistic, and values-oriented, skills-based living. While exploring how AI and automation will affect our future, we can improve our present by exploring how positive psychology is improving humanity today.

- **Happinomics will become a mandated measure.** The Internet of Things (regarding objects) and the Internet of Pings (regarding people's personal data) will provide multiple layers of accountability for society to deal with. Conversations need to move beyond privacy around this data to create comprehensive ways people's identities and actions will register within the public domain. Happiness or well-being, as measured by wearable or other devices, can be utilized for such a purpose and can be correlated with global measures such as gross national happiness and the genuine progress indicator. In this way, countries can perceive in real time if policies are helping or hindering the general welfare of their citizens.
- **Big problems call for BIG solutions.** A basic income guarantee may not work in the U.S. But the growing wealth gap, driven in part by widespread adoption of automation technology, demands workable short- and long-term solutions regarding employment and income. If millions of Americans and global citizens will soon be out of jobs because of AI or automation, we need to test pragmatic ideas now to deal with this issue.

eleven | The Evolution of Economics

Fall 2030

"Hi, John. Time to wake up."

I shifted in bed and nodded to Caffie, my personal robot assistant. She was modeled after the personal robot created by the company Robotbase[1] in 2015. The device featured an oval "head" that functioned like an iPad and showed the face of an animated young woman with large blue eyes. Barbara and I nicknamed her Caffie because she got us going in the morning like our caffeine.

"Barbara went to work," said Caffie as I sat up. "She left you a short video. Do you want to see it?"

"Yes, please. Thanks."

Barbara appeared in place of Caffie's face on the screen. "Hi, honey. I'm meeting with Susan to go chat with a potential wedding client. They apparently have a huge budget, so we're pretty psyched. Should be home around lunchtime." Caffie's face reappeared. "Would you like some fresh coffee? I can also turn the heat up downstairs if you like," she offered.

"Sure, all the usual stuff would be great, thanks." Caffie was able to communicate with our smart home. She was essentially a really clever user interface that featured speech and facial recognition and converted it to code or actions the devices in our house could understand. This was a very good thing, as asking someone to write code in order to make a cup of coffee in the morning would be a form of cruelty.

I used the bathroom while Caffie stayed outside in the hall. While a lot of younger people apparently didn't care if their robots watched them pee, I still felt weird about it. Anthropomorphizing is a pretty personal process in that way. We did have a smart toilet,[2] however, so if there were any potential health issues based on my morning constitutional, Caffie would let me know later in the day. Typically she didn't tell me what might be going on with my digestion but just added extra vitamins to a glass of water I would drink or whatever I needed to help me out.

I stopped quantifying my physical health long ago, at least on a daily basis. I'd spent enough time figuring out how to optimize my diet that Caffie just alerted me when I was out of balance. She also helped me control my diet. This included things such as locking the cabinet where we kept our blue chips or showing me photos of how I looked when I was at my heaviest weight. I'd programmed her to do these things, of course, so as Big Brother as it may sound, she was simply a manifestation of my preferences. Granted, she still pissed me off when she denied me chips during a sci-fi movie marathon late at night. But she was essentially me, so I was actually just getting mad at myself.

Sounds complicated, but you get used to it.

A lot of the technological advancements that allowed programs like Caffie to flourish traced their roots to a company called Wit.ai,[3] whose focus was on creating natural language commands for the Internet of Things. This was extremely smart, as it could help programmers translate human language into code that devices could understand. Facebook purchased the company in 2014, so now you could post to your friends by just speaking and using basic commands. Saying phrases such as "Post photo" meant people had done a lot less typing over the past number of years. I'd even lost my thumb texting calluses from 2017.

The following diagram from Wit.ai's[4] site shows you how their service works:

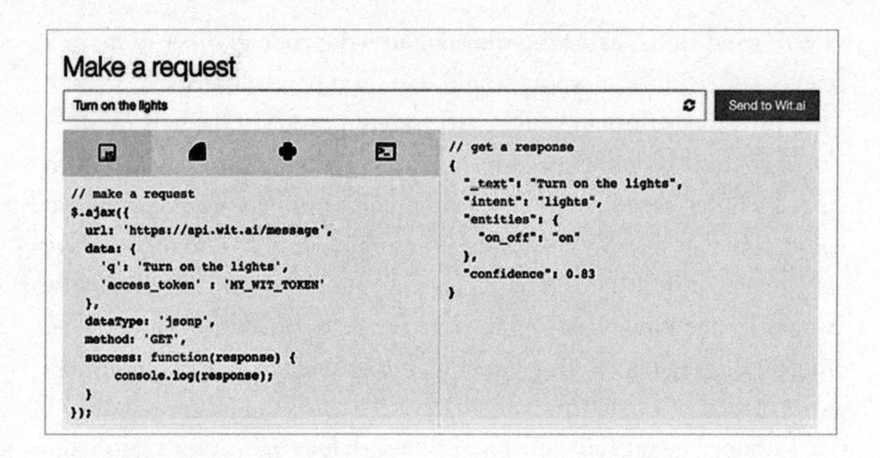

Cool, huh? While the service started out for developers to build applications, it became simple enough for average consumers to create their own apps as well. Wit.ai also "learns human language from every interaction, and leverages the community; what's learned is shared across developers."[5] This aggregate learning about language was used to help create ethical standards for all AI systems. Ethicists working with programmers and the general public realized it was a lot simpler to analyze how people really wanted to shape their lives regarding the Internet of Things than to create scenarios based on guesswork. Fortunately Mark Zuckerberg had been open to this use of the company's intellectual property, as he saw that common ethical standards also provided a new revenue stream for Facebook. When people utilized the service in ways that reflected their values, advertisers could transparently target users with greater relevance, and they paid Facebook a cut of their revenue. Everybody won.

This concept came to be known as Values by Design, based on the policy framework known as Privacy by Design (PbD). There are variations on how PbD works, but its foundational principles[6] provide ways for individuals to choose how they control their personal data. For this to happen, privacy measures have to be embedded into the design of a program before it's provided to the public. The program also has to allow for transparency and be user-centric, so an individual fully understands how and where his or her data is being used in a transaction. The difficulty with most regulation regarding privacy is that it has to deal with a person's data *after* it's already been tracked or sold. Allowing individuals to know how their data might be used before it's tracked puts the power in their hands to control their identity. They can still interact with advertisers, brands, or anyone else as they see fit. PbD just means they have a handle on who wants to interact with them and why, and provides a framework for them to do so.

Values by Design (VbD) provides the same kind of framework for individuals, but adds a layer of contextual ethical information to standard privacy concerns regarding people's interactions. For instance, my friend David works in marketing and public relations and does a lot of sharing via social networks. He's very savvy about his personal data, and protects it via a personal cloud. This way, advertisers and brands know how and when to reach out to him for any products he may want to buy. For this side of David's life, Privacy by Design meets his needs. However, Dave's spirituality is also a big part of his life. He's practiced in the art of meditation and runs classes in a school for people to explore faith-based purpose in their lives. For the parts of his life focused on these ideals, Values by Design provides a much more relevant way for people to reach out to him than with product pitches alone. VbD also helps certain companies know they *shouldn't* reach out to David, such as alcohol brands, since he doesn't drink based on his beliefs. In this way VbD lets people curate their data to be used to track not just general demographics about their lives (age, gender, location), but the deeper ethical and moral beliefs that comprise their values.

The Values by Design mind-set extends into the physical world as well, informed by people's personal data and digital habits. For instance, if a person valued the environment, he could conserve electricity in his home to be measured in his local electric grid and benefit his community. Whereas people's values had been indirectly measured via metrics such as census or survey data, now information was much more granular because people's actions were measured by the sensors in their phones and the world around them.

"Do you want to listen to the news?" Caffie asked me once I'd left the bathroom.

"Sure," I said, and she turned on NPR. I was at the top of the stairs, so I gently grabbed Caffie and brought her down to the kitchen. While years ago engineers at Honda had created the ASIMO robot,[7] which could walk and navigate stairs, it still cost more to get those types of robots than ones with wheels, like Caffie. Besides, I kind of liked that she was dependent on me for something.

After I finished my coffee and breakfast, Caffie's face came back on the screen, replacing the radio icon. "Did you want to hear about the neighborhood's gross national happiness statistics?"

"Yes, please," I answered.

"To begin with, life satisfaction scores overall have increased in the past twenty-four hours," said Caffie. "I checked people's social media posts and this appears to be due to the increase in temperature yesterday." Caffie utilized a methodology for these results created by researchers at the University of Pennsylvania as part of their World Well-being Project.[8] The researchers had discovered that the well-being of a community can be determined through the collected posts of its individuals and that the information derived from the data has practical applications across a broad range of disciplines, from marketing to medicine to national security.[9] Other companies had done this for a while—Sickweather,[10] for example, distilled social media posts about illness, which was extremely helpful while our kids were still in school. We'd know with hyper-local accuracy if we should keep our kids home to prevent them from getting the latest round of flu or whatever was going around.

"Hmm," Caffie continued. "Environmental scores are down."

"I bet it's because Little League started up, right? Excited parents forgot to dispose of their water bottles or whatever?" When I said "water" an icon on my smart fridge lit up with a small ping. If I said "pour" it would dispense water in a glass but I ignored it.

"Exactly," said Caffie. "I just accessed the town's maintenance filings and they reported a higher level of trash than usual."

"Will they get fined? The parents?"

Caffie's animated head bobbed up and down. "Yes, since they were in a public park. Their GPS logs can be cross-referenced with their bottle purchases and amount of water they ingested, via their wearable devices."

"But that's only if they opted in to that environmental values initiative, correct?" The coffee smell still hung in the air, so I started making another pot. Caffie knew not to ask me to let her do it—I enjoyed the ritual of making coffee even though I'd given the right of first pot to her.

"Yes," said Caffie. "It's their equivalent of locking blue chips in a cabinet. The program they signed up for only fines them for behavior they want to prevent in the future. It's actually a win-win. When they remember to recycle, everyone is happy. When they forget, they're irritated at the fine, but their money goes to clean up the environment."

"And the people who freak out about big government or think this technology is invasive?" I asked.

"They probably haven't opted in to these programs." Caffie smiled. "And don't have personal robots."

"How are the psychological well-being scores for the neighborhood today?" I asked. These metrics referred to the emotional health of people in the community who opted in to a program that allowed facial and biometric information to be made public to other participants.

"Harry seemed lonely yesterday," said Caffie. Harry was a widower who lived three doors down from me. He walked his golden retriever religiously every morning and afternoon. Whenever I saw him I made a point of yelling, "Hi, Harry!" and he'd give me a vigorous wave before walking away.

"How can you tell?" I asked. "Normally he just seems determined to me. His dog pulls him pretty hard, so he may just be focusing on not falling over."

"I'm equipped with pupil-level analysis. To your point, his face registered physical exertion. But his pupils contracted in a way that correlates to melancholy."

"Bummer," I said. "Has he posted any needs to the community in the past few days? Something at the store I could pick up for him? I'm assuming he's not a big Oculus Rift user or I could try and chat with him virtually."

"He hasn't posted any needs, and he's not active virtually," Caffie agreed. "But I have a suggestion, based on my deepest cloud-based artificial intelligence, accessing all the world's information in nanoseconds."

That sounded ominous.

"What's that?" I asked.

"He just left his house to walk his dog." Caffie winked. "In terms of what he might need, why don't you just ask him yourself?"

Purpose Versus Purchase

Sung to the tune of "If I Only Had a Brain."

When your outlook is financial, increasing growth seems quite substantial, and only money plays a part.

Economic-ally speaking, sorry but your theory's leaking, you've got to measure heart.

I know at first it sounds deceiving, that money can't increase well-being, but it's time for a clean start.

Please be open and not churlish, there's more metrics make you flourish, when you start to measure heart.

Oh I can see just why, you're struggling to comprehend. But don't worry now, my data-loving friend. Here's where we start, it's not the end.

So don't fill yourself with tension, 'bout these metrics that I mention, like education, health, and art.

'Cause the paradigm is changing, economics rearranging, now we get to measure heart.

—WRITTEN BY JOHN C. HAVENS FOR THE GROSS NATIONAL HAPPINESS CONFERENCE, 2014[11]

I realize my opening narrative may seem a bit creepy. Caffie and her data are wildly invasive, and for people who don't like the idea of the government or other people being up in their grill, this future scenario may cause concern.

Fair enough.

But as a reminder, all the tracking examples I provided are or could be taking place with you and your data right now. You just don't know about them. Picture dozens of Caffies online and in the world around you via the Internet of Things all gaining information about your identity without your consent or knowledge. None of the data collected directly benefits you. Or if it does, you may not know about it because nobody asked you your opinion on the subject. Is that a better scenario?

After doing so much research about artificial intelligence and emerging technology, I've realized that being human in the near future is going to be an acquired skill. The tools, apps, and devices we use today will only grow to encompass more of our consciousness.

Maybe the power won't shut down that often, or batteries will never run out. Maybe Wi-Fi will be ubiquitous, so people won't have to worry that their emotional recognition tools or biometric sensors will be unavailable. Perhaps interacting with another human without any tech will seem as outdated to someone as learning ancient Latin might be today.

Well, that would be weird.

But data about our well-being, if transparent to the public, will provide an accountability for humanity as I've described in my scenario with Caffie. While it may be simpler to avoid dealing with other people by escaping into a virtual or augmented reality of some kind, our hearts and minds will always yearn for human contact along with the benefits technology can bring.

However, a risk we must face is that technology could make it easier for us to avoid those in need. I've written before about augmented reality contact lenses being programmed to keep users from seeing homeless people or to avoid any news mentioning unpleasant subjects. We already exhibit this kind of behavior via low-tech means. But if machines take our jobs, or do a great deal of the thinking we used to be challenged to do, we're going to crave the natural boost that helping others provides.

Maybe drug dealers in the future will be people running nonprofit organizations, offering select individuals the opportunity to increase their self-esteem through volunteering. Perhaps the most popular video games will allow users to teach needy kids, or solve problems such as world hunger.

That would also be weird.

But don't you want to be challenged? Don't you yearn for someone to recognize the skills and talents you have so you can live to your full potential in the world? I do. The exciting news is that a great deal of emerging technology, including AI, can help us recognize how our talents affect those around us. But Big Data unleashed in a world based on consumption is incredibly limiting. It demands that we focus on ourselves above others. We earn money to be able to consume products. We consume media for our own entertainment and to learn about products we'll want to buy.

But in the wake of AI and automation we may not be able to earn enough money to consume products. We won't be able to afford the entertainment needed to escape lives devoid of meaning without work. So what are our choices if widespread automation comes to pass?

1. **Despair.** Always an option, available in abundance.
2. **Pursuit of pleasure.** Attractive, but finite. And pleasure is often defined by the pain we've escaped.
3. **Pursuit of purpose.** Bingo.

Pleasure is like happiness. You can pursue it, but it spikes and falls and often leaves a vacancy of meaning in its wake. Purpose provides an ongoing journey. Purpose provides a way to increase intrinsic well-being for the rest of your life. And it can be pursued with the aid of technology, or without. The power to turn it on or off is up to you.

When Data Gets Happy

In *Hacking H(app)iness* I described in great detail how the concept of gross national happiness (GNH) was created. Inspired by a speech delivered by

Robert Kennedy, the country of Bhutan's fourth dragon king, Jigme Singye Wangchuck, coined the term *GNH.* He felt the GDP didn't serve as an accurate measure of value for his country based on their Buddhist spiritual values. His colleague Karma Ura then created the Centre for Bhutan Studies, along with a survey tool that measured Bhutan's well-being via a list of metrics focused on measures beyond the GDP's largely financial focus.

Here's an example of these GNH metrics as created by the nonprofit organization the Happiness Alliance.[12] I'm on their board, and their founder Laura Musikanski's work is what taught me about the nature of measuring well-being for policy.

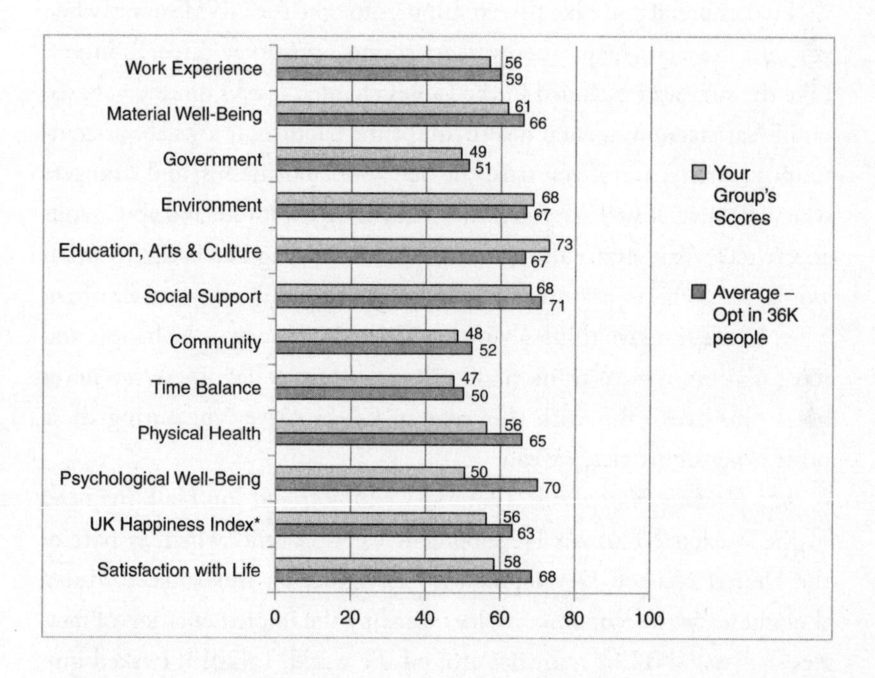

This data is based on the GNH index survey[13] Laura's organization created. It's a tool developed to model Bhutan's metrics and measurements, as Laura described in her paper "Happiness in Public Policy," for Walden University:

The happiness movement represents a new paradigm where social, economic, and environmental systems are structured to encourage human well-being in a sustainable environment. Bhutan has adopted Gross National Happiness (GNH) as a way of determining its society's success in contrast to purely economic goals and the singular use of the gross domestic product indicator. Bhutanese policy promulgation includes use of a GNH screening tool. In the United Kingdom, happiness indicators are being used to collect data and the government is starting to explore their application to policy. The Bhutanese GNH policy screening tool has been adapted for the grassroots activists, providing opportunities for everyone to participate in the happiness movement.[14]

I recommend you take fifteen minutes to take the GNH survey when you can. It's here: http://survey.happycounts.org/survey/directToSurvey. Like the surveys I included in the values chapter, it asks questions based on life satisfaction, mental health, and time balance. It asks about community vitality, and if you trust the neighbors, businesses, and strangers where you live. It will ask you how often you've felt loved, and about your access to artistic and cultural experiences on a regular basis. In other words, it's going to ask questions you probably don't ask yourself often. Most of the time we think about how to make money or be happy, and since it's common to think making money brings happiness, we never break this cycle. But with this type of GNH survey measuring these other types of metrics, we can.

For *Hacking H(app)iness*, I interviewed my friend Jon Hall, the head of the National Human Development Reports unit, which is part of the United Nations Development Programme, on this subject. He's a thought leader in economic circles regarding the implementation of metrics such as GNH in countries around the world. In 2013, I asked him where he felt the Beyond GDP movement (encompassing metrics such as GNH and the genuine progress indicator, which I describe below) could evolve in the future. Here's what he said:

In five years' time I think people will be using this type of data to implement policy. In twenty years this could be very radical. Well-being could actually change the way that the machinery of government is put together. We'd have a realignment of how different ministries work together and how decisions are made. It changes everything.[15]

Jon also contributed to a recent World Happiness Report,[16] created in conjunction with the United Nations, which was edited by leading economists John Helliwell, Richard Layard, and Jeffrey Sachs. Reports like these and surveys like Laura's are not intended to simply measure mood but to inspire genuine change. Measuring well-being is not a passing trend, but an entrenched methodology.

In this regard, my opening narrative is designed to show how this type of data could integrate with our lives in the future via artificial intelligence and Big Data. This is where I think AI and emerging tech will be most beneficial—where we can know how we're projecting our identity to the world so we can give and get all we need to increase well-being without money. It may sound complicated, but so is replicating the human mind and imbuing machines with consciousness.

The genuine progress indicator (GPI) provides another set of metrics such as GNH to move beyond GDP and gain a more holistic measure of well-being. Currently the GPI has been adopted in a few states in the U.S., including Maryland and Vermont. Here's how GPI is described on the Maryland State website:

The Genuine Progress Indicator (GPI) provides citizens and policymakers fruitful insight by recognizing economic activity that diminishes both natural and social capital. Further, the GPI is designed to measure sustainable economic welfare rather than economic activity alone. To accomplish this, the GPI uses three simple underlying principles for its methodology:

- *account for income inequality,*
- *include non-market benefits that are not included in Gross Domestic Product, and*
- *identify and deduct bads such as environmental degradation, human health effects, and loss of leisure time.*

The GPI developers identified 26 indicators, then populated them with verifiable data. As one example, the pure economic activity stemming from the explosive growth of urban sprawl contributes greatly to the GDP. Yet, along with sprawl come non-economic costs such as increased commuting time, increased traffic congestion, land use conversion, and automobile impacts. In short, just because we are exchanging money within an economy does not necessarily mean that we are sustainable or prosperous.[17]

I wrote an article about GPI for the *Guardian*[18] in which I show how the GPI factors in something the GDP doesn't: a central tenet of business called double-entry bookkeeping. For instance, if you own a retail store, you can't base your success on your gross profits. You may sell all your products, but you have to account for paying your employees and overhead before figuring out your net or actual revenue. Shockingly, the GDP doesn't utilize this methodology, but focuses on metrics regarding growth alone to base its results. So as the preceding quote describes, a city's GDP may increase when people move and get jobs there. But those people also decimate the environment and air quality from sitting in traffic on their commute. These degradations impact people's lives negatively, increasing health care costs or taxes once damage is already done.

"Message number one from GPI: There's good growth and bad growth," noted Marta Ceroni, Ph.D.,[19] during our interview for this book. Marta is the executive director of the Donella Meadows Institute,[20] in Vermont, and her mission is to align economics with the reality of a finite planet using systems thinking and collaborative learning. "In terms of the GDP, growth is always seen to be good—even with something like a hurricane striking a region that provides jobs to help clean up after the disaster. But in what way could a hurricane be said to *increase people's well-being*? Growth in and of itself is an incomplete metric along these lines."

Another pivotal aspect of the GPI is that it's designed to measure things that don't normally have a price within the GDP. In that sense, it's somewhat of an intermediary measure between the GDP and something like GNH. As Ceroni points out, "We're making sure that these efforts around GPI map onto a vision that is shared and well understood as opposed to having sparse indicators that give a general idea but without specific intentionality." One very powerful area along these lines is the example of income inequality, a metric that's currently not measured in terms of these types of global indicators in any way except with the GPI. This factor means GPI offers much richer data surrounding issues of women, minorities, and those with disabilities than the GDP. It also helps put monetary numbers to efforts of men and women staying at home to take care of their children versus working. For instance, do stay-at-home parents' "jobs" count toward increasing the well-being of their community or the world at large?

Maryland says yes, at least in regard to measuring the "value of housework" in their genuine progress indicator:

> *The housework performed by individuals, families, and households is essential to the functioning of the State's economy. Unpaid housework, parenting, and other activities performed in the home are required to support individuals' economic activities, but are uncounted in standard measures of economic activity. . . . Finding an accurate means to measure housework can also be critical to understanding whether the critical needs of our children and most vulnerable adults are being met, how we are coping with increasing demands on our time from work, and how much of our income is devoted to repairs to our homes or vehicles when routine maintenance needs go unmet.*[21]

To help calculate household income, the government uses something called the American Time Use Survey[22] (ATUS), which measures a number of metrics, including child care, volunteering, and socializing. Utilizing data from ATUS, the state then compares data for Maryland housekeeping professionals against how much time people spend doing unpaid housework. The logic here is that if people weren't doing their own housework, they'd need to pay someone else to do so. Here's the final formula Maryland uses to make its calculations:

(yearly hours spent by Marylanders on housework) x (wage rate for housekeeping workers)[23]

Imagine how much more granular this type of data will become when it's aided by sensors, social networks, and the Internet of Things. A great deal of the guesswork and bias of surveys will be replaced by real-time monitoring of our actions and intentions. Combined with data mirroring our preferences or values, my opening narrative is far from unrealistic.

From Money to Meaning

One of my favorite books is by E. F. Schumacher, a German Rhodes Scholar in economics who taught at Oxford University. It's called *Small Is Beautiful: Economics as if People Mattered*,[24] first published in 1973. The book is a paradigm-shifting read that warns against the dangers of finite "scientific" methodologies championing growth as the ultimate aim of economics. The *Times Literary Supplement* ranked the book among the one hundred most influential books published since World War II, and Schumacher's critique of Western economics has also deeply influenced modern thought on sustainability and environmental issues today. Here's an excerpt written by Theodore Roszak in the introduction of the book to give a sample of Schumacher's thinking:

> *What sort of science is it that must, for the sake of its predictive success, hope and pray that people will never be their better selves, but always be greedy social idiots with nothing finer to do than getting and spending, getting and spending? It is as Schumacher tells us: "when the available 'spiritual space' is not filled by some higher motivations, then it will necessarily be filled by something lower—by the small, mean, calculating attitude to life which is rationalized in the economic calculus." We need a nobler economics that is not afraid to discuss spirit and conscience, moral purpose, and the meaning of life, an economics that aims to educate and elevate people, not merely to measure their low-grade behavior.*[25]

Now that we're able to measure emotion and well-being via sensors and data, isn't it time to evolve the GDP? Metrics included in the GPI provide a great start to this process, since it's a complement to GDP and expands a sense of values for citizens beyond financial measures. This would actually support the mind-set of the creators of the GDP. As Sankaran Krishna points out in his article "The Great Number Fetish": "Simon Kuznets and John Maynard Keynes, both pioneers in its creation and measurement, warned against confusing GDP with anything other than a measure of the sum of economic activity of a society, and especially against confounding it with societal welfare."[26]

Adam Smith, the father of modern economics, is mostly quoted regarding his book *The Wealth of Nations*, which describes his famous notion of "the invisible hand," in which he claims people's efforts to maximize personal gain in a free market benefit society. But recent scholars have begun to note, as with Simon Kuznets, that Smith's ideas along these lines may have been given too great a priority. As it turns out, Smith wrote another book, *The Theory of Moral Sentiments*,[27] in which he said, "How selfish soever man may be supposed, there are evidently some principles in his nature which interest him in the fortune of others and render their happiness necessary to him though he derives nothing from it except the pleasure of seeing it."

While the ideas in Smith's two books may seem contradictory, they're in fact complementary. Money will always play an important role in our lives, as a standard of value comparison or to pay for basic needs. But regarding moral sentiments, Smith talked a great deal about the idea of "the Impartial Spectator." This figure represents a form of our conscience that motivates us to do good for others so we can be perceived as "lovely," or lovable. Similar to the nature of altruism, in which our self-esteem is increased by helping others, being "lovely" in society ensures we'll get loved in return.

How would the world look if the fortune of others were prioritized over growing wealth for ourselves?

How would the world look if maximized personal gain were measured in the happiness we create?

It would look amazing.

So why have we focused so much on Smith's ideas in a financial context? I interviewed Russell Roberts,[28] a John and Jean De Nault research fellow at Stanford University's Hoover Institution and host of the popular podcast *EconTalk*, to ask his thoughts on this question. Roberts recently completed a book about Adam Smith's *Theory of Moral Sentiments* called *How Adam Smith Can Change Your Life: An Unexpected Guide to Human Nature and Happiness.*[29] Here's what Russell had to say regarding the GDP and its focus:

> *I'm not a big fan of monetary measures of well-being. I concede they matter, they're just not decisive. As a good economist I understand that money alone doesn't give me happiness. But I also think we overstate the shortcomings of GDP. The bigger problem is psychological, not governmental. We as human beings tend to focus too much on the financial aspects of life versus the less tangible things that give life meaning.*

These less tangible things that give life meaning are easier to measure than ever before. And we don't have to quantify our emotions and actions for the rest of our lives to figure out what those things are. We just need to shift our focus from cash to caring, increasing our intrinsic well-being by helping others do the same.

Here are the primary ideas from this chapter:

- **Citizen data.** In the near future our actions and personal identity will reflect our political and ethical ideas to our real and virtual communities. While this data will provide accountability for individuals, it will also encourage transparency for government. Accountability based influence will help shift politics when our data can lobby for our interests in real time.
- **The GDP is DOA.** While there's no need to mock the GDP or the benefits it has provided, we must complement or replace it with more modern metrics. We can't tout the benefits of Big

Data for business while ignoring the vast stores of information now available via sensors reflecting emotion, health, and mental well-being.

- **The virtual impartial spectator.** If Adam Smith were alive today, his notion of the impartial spectator would certainly factor into people's actions as reflected by social networks and the Internet of Things. In the same way GDP should shift to measure well-being versus wealth, our lives should measure our positive impact on others versus only our online influence.

twelve | Our Genuine Challenge

Special Book Feature: Choose Your Own AI-dventure!

You are reading this book, wondering why the author misspelled *adventure*. Then you realize he's provided you the chance to create your *own narrative* about artificial intelligence via the Choose Your Own Adventure books format he loved so much as a kid.

Present Day

You put down your copy of *Heartificial Intelligence*, reflecting on the nature of artificial intelligence. While you appreciate the notion that tracking your values could be beneficial, you think AI concerns only geeks and academics. Picking up your phone, you troll your Facebook feed for the fourth time today and watch a video featuring cute pandas going down a slide after you accidentally click on an ad for a weight loss supplement. Later in the afternoon, while checking your e-mail, you get an invitation to connect with someone you don't recognize on LinkedIn. When you click on her profile you see you don't share any personal connections, but she does work in the weight loss industry, making you wonder if she contacted you because of the ad you clicked on earlier in the day.

- If you're freaked out about being tracked from Facebook to LinkedIn, select CHOICE #1 below.

- If you delete the e-mail, irritated by the ad but not concerned, select CHOICE #2.
- If you buy the weight loss supplement, select CHOICE #3.

CHOICE #1

While you've been aware that organizations track you online, you've never made the connection of how intimately your everyday actions are being monitored. You become irritated at the fact that the sum total of your digital day amounts to forty-five seconds of panda-rific glee and the uninvited pressure of body image issues. You decide to learn more about how to control your personal data and visit Personal.com to learn more.

- If you decide to try controlling your personal data, select CHOICE #4 below.
- If you don't sign up for Personal.com or a different cloud provider, select CHOICE #5.

CHOICE #2

For the following three to four weeks, you receive multiple ads for the weight loss supplement you never wanted in the first place. Flustered at this mindless repetition, you burn more calories in angry frustration than if you'd taken the supplements. Unfortunately, your new wearable device also notes your stress has increased to toxic levels, and you die.*

*Remember how the Choose Your Own Adventure books would do this? You'd turn the page after choosing to climb a tree or take a right turn or whatever, and they'd just say, "You're dead!" That shocked me the first few times it happened, but then I just assumed the writer created those short endings because he got a phone call or wanted to watch *Diff'rent Strokes*.

CHOICE #3

After clicking to purchase the weight loss supplement, you continue to receive multiple ads for the same supplement you just ordered for the remainder of your life. While you try to counter the stress of waiting for the supplements to arrive in the mail, you consume more food in a two-week period than you have in the past four months. While straining to wedge a fresh slice of turducken into your gaping maw you choke to death on a stray bone and the irony of your untimely demise.

- If you decided to place your consciousness into a mindclone before you died, select CHOICE #6 below.
- If you have a personal robot assistant in your home in the midst of this crisis, select CHOICE #7.

CHOICE #4

You decide to use Personal.com and begin to understand how many other organizations access your data. Realizing how simple it is to track your intentions and actions, you begin wondering how conversations about privacy can move toward issues of control and ethics as we approach the future.

- If you're wondering if the author has any financial ties to Personal.com, select CHOICE #8 below.
- If you decide to sign the Future of Life Institute's petition regarding beneficial AI, select CHOICE #9.
- If you think ten choices is probably enough for this Choose Your Own Adventure format, select CHOICE #10.

CHOICE #5

Ignoring your momentary interest in protecting or controlling your personal data, you don't sign up for Personal.com and instead focus your energies on preparing for your blind date later in the evening with a guy you met on OKCupid. Excited at first by his rugged good looks and natural charm, you end up choking on the lime in your vodka tonic when he "randomly" mentions weight loss supplements, and you die.

CHOICE #6

While waiting the interminable amount of time it takes for your weight loss supplements to arrive, you decide to sign up at LifeNaut.com, as discussed in *Heartificial Intelligence*. It's actually a very gratifying experience, and you realize how much more you appreciate uploading your photos and videos to a site focused on helping you memorialize what you hold most dear versus one that simply facilitates a constant stream of cultural ephemera. While you don't feel you're uploading your consciousness per se, you do appreciate the ability to curate the aspects of your personality for other loved ones to enjoy in the future.

CHOICE #7

While you enjoy the companionship of your personal robot, you become irritated at how intimate it's becoming with the rest of your house. After your smart air conditioner refuses to turn on during peak usage hours to support the local electric grid, you attempt to shut off your robot's user interface to regain control of your house, at which point the robot kills you and you die. When the police arrive to investigate the crime, your robot casually brings up the idea of weight loss supplements to an officer it has identified as overweight via his facial recognition scan. He takes your robot home after ordering the supplements and eventually offends his smart fridge after complaining his Ben & Jerry's has freezer burn, at which point the robot kills him to defend the fridge's honor. Keanu Reeves ends up starring in and producing a movie about the incidents, which is sponsored by a prominent weight loss supplement company.

CHOICE #8

I don't have any financial ties to Personal.com. I've interviewed some people from the company and just think they're really smart. When people ask me how they can protect or control their personal data today, I always point them to Personal.com because they (a) know a lot more than I do, and (b) don't deal in weight loss supplements, to the best of my knowledge.

CHOICE #9

Excellent choice. You should also read the Research Priorities document at: http://futureoflife.org/misc/open_letter. While you may think AI is only for geeks, that's not the case. Big Data is worthless without deep learning techniques or other methodologies that can make sense of the noise. The algorithms already powering the Internet comprise the beginnings of AI in every manifestation we may see in the future. Don't worry so much about *The Terminator* scenarios—focus instead on supporting the AI programmers, ethicists, and experts from multiple disciplines trying to codify rules for how humans and machines can live in peace and prosperity, hopefully without the need for weight loss supplements.

CHOICE #10

Agreed. Please also note I have no issues with any of the good people working at the 17,428 weight loss supplement companies flourishing today. My point in using them as an example was simply to indicate how what we click affects what people think we value, which can follow us the rest of our lives.*

*I am open to receiving funding or sponsorship from any weight loss supplement company, particularly if your product comes in turducken flavor.

> *We'll spend the next decade—indeed, perhaps the next century—in a permanent identity crisis, constantly asking ourselves what humans are for. In the grandest irony of all, the greatest benefit of an everyday, utilitarian AI will not be increased productivity or an economics of abundance or a new way of doing science—although all those will happen. The greatest benefit of the arrival of artificial intelligence is that AIs will help define humanity. We need AIs to tell us who we are.*[1]
>
> —KEVIN KELLY, "THE THREE BREAKTHROUGHS THAT HAVE FINALLY UNLEASHED AI ON THE WORLD"

I agree with Kelly that AIs will help define humanity. But I disagree that we need them to tell us who we are. The algorithms and learning that AI provide may bring intellectual enlightenment, but it's up to us humans to foster our own transformation.

That's not pejorative toward AI, by the way. It's just that AIs are not people. At least not yet.

But you are. You've got a brain, and a mind. The lump of tissue inside your skull contains consciousness as well as neurons. Maybe this consciousness is a soul, maybe it's a basic form of cognition. But it's at your disposal right now.

You've also got a heart. Not just the physical pump, but the emotional life created by your neuron firings, hormonal releases, and physiological manifestations of well-being you share with the world.

Plus you've got your values. These include universal codes about not killing or stealing from others, caring for your family, and watching out for your neighbors.

Today, machines don't have minds. They don't have hearts. They don't have values. But they're programmed by people who do. While you may not be able to directly influence the outcome of AI research or ethics, your vote still counts about how we move toward the future as humans. You're allowed to be angry, scared, excited, or indifferent about AI and its potential outcomes. But the technologies speeding AI forward are advancing rapidly however you feel. Thankfully a great deal of people in and around the AI community realize why ethics should play such a big role in its development. But you can help as well. While it may take experts to program the algorithms and systems driving our machines, you're allowed to decode what makes you tick as a human.

My research and life experience creating *Heartificial Intelligence* has led me to believe this journey of identifying your humanity stems from tracking your values. It's a simple process at first—you don't have to lose weight or try to be happy to reap its benefits. Just ask yourself what specific ideals you wish to follow in your life and see if you're spending time with them every day. You'll likely discover you're out of balance with a few factors guiding your life. If you're like most people, you spend so much time at work that your family life or time for learning is diminished. Or your health has gone downhill because you spend too much time glued to a screen.

So take those insights, and switch up how you spend your day for a few weeks. See how changing activities based on your data increases your well-being and the happiness of the people in your life.

This is the type of Big Data we should use to guide AI ethics and human transformation for the future.

Chapter Review Points

Here are the key points from each chapter of the book—important takeaways to help you live a genuine life as we face the effects of artificial intelligence:

CHAPTER ONE: A BRIEF STAY IN THE UNCANNY VALLEY

- **Our happiness is being defined by how we are tracked.** There are two ways to deal with this:
  - ✧ *Continue utilizing the current model of aggressive, clandestine surveillance.* Algorithms and data brokers know more about

our lives than we do. Our happiness or well-being is measured only within the context of what we purchase based on our on- and off-line behavior.

    - *Create a new model featuring an environment of trust, where all parties in a transaction are accountable for their actions.* Commerce can flourish within an environment where people have transparent access to their data and can best reflect on their well-being.

- **The uncanny valley of advertising won't last very long.** As preferential algorithms improve, if we continue in the direction we've been going, we'll stop seeing the traces of how companies are tracking our lives. In the same way we've given up control of our personal data, we'll lose the opportunity to understand the logic of how people are manipulating and affecting our well-being.
- **Accountability based influence will provide technological transparency.** Whether we like it or not, our tracked actions will be broadcast to the people in our lives in ways we've never experienced before. This exposure will inspire individuals to better control their personal data, while also providing opportunities for more insightful self-examination in the wake of technological overkill.

CHAPTER TWO: THE ROAD TO REDUNDANCY

- **Human capability is finite in nature.** While heady arguments take place about the nature of sentience in machines, there's no disputing that Kiva robots in warehouses function far more efficiently than humans. They move across football-field-size buildings at a rapid pace without ever needing a break, overtime pay, or health insurance. Jobs in fields such as legal processing and medical imaging face the same dreary outlook when compared to the analytical prowess of computers. We're building the machines that are replacing most, if not all, of humanity's jobs and we've misused time focusing on *when* specific verticals will become automated versus *what to do when they are.*

- **People need to get paid.** As much as I support the growing sharing economy and other models I elaborate on later in the book, I don't see consumerism or capitalism going away anytime soon. A seminal fact within economics is that for markets to be sustainable, consumers need to be able to afford the items created by producers. So any solution regarding automation that includes utopian visions of displaced workers getting to pursue new interests has to account for this undeniable fact.
- **People need purpose.** There's a rising trend to encourage happiness and well-being in the workplace. Much of this work is centered around helping people identify what gives them a sense of "flow," or what activities bring a deep sense of meaning to their lives. While working in a factory or for UPS may fulfill a sense of purpose for employees under normal conditions, the examples I provided in this chapter hardly qualify as environments where a human can thrive.

CHAPTER THREE: THE DECEPTION CONNECTION

- **Artificial intelligence replicates but can't replace.** It's illogical to think we can copy ourselves or loved ones without the algorithms representing us evolving into unique personalities. This means that on top of avoiding the pain and growth associated with loss, if we're able to mimic people's consciousness in the future it may end up looking very different from those we knew.
- **Artificial intelligence is biased by anthropomorphism.** Just because we may be tricked into thinking something is real, it doesn't mean it actually is. While I can respect a person's right to believe his or her autonomous car is alive, for instance, we still need laws governing the culpability of these vehicles in relation to the humans affected by them.

- **Artificial intelligence may hurt our ability to help.** Emotionally oriented AI programs like the one driving the robot Pepper are ostensibly designed to help us. But in providing easy companionship their cloud-driven technology may rob us of our ability to learn empathy.

CHAPTER FOUR: MYTHED OPPORTUNITIES

- **Advertising-driven algorithms lead to nonsense.** The nonsense is both technical and metaphorical. Beyond issues of human error creating the algorithms in these systems, most programs can be easily hacked. Data brokers sell our information to the highest bidder, and the entire system is predicated on purchase versus purpose. If humanity is to be eradicated by machines, let's not have it be a market-driven massacre.
- **Ethical standards should come before existential risk.** A majority of AI today is driven at an accelerated pace because it *can* be built before we decide if it *should* be. Programmers and scientists have to be held to moral standards along with financial incentives throughout the AI industry, *today*. The Future of Life Institute's petition provides an excellent starting point for conversations along these lines.
- **Values are key to vision.** Counterintuitive or not, human values need to be baked into the core levels of AI systems to control their potential for harm. There are no easy workarounds. Asimov's fictional laws of robotics or well-intentioned myths like Google's outdated mission statement need to be replaced by pragmatic, scalable solutions.

CHAPTER FIVE: ETHICS OF EPIC PROPORTIONS

- **Robots don't have inherent morals.** At least, not yet. It's imperative to remember that programmers and systems need to implement ethical standards from the operating system level on up. Otherwise a properly operating algorithm simply seeking to fulfill its goals may pursue a course of harmful action.

- **Organizations need AI accountability.** P. W. Singer's idea of a "human impact assessment" provides an excellent model for society to pursue regarding organizations creating or utilizing AI. In the same way companies are responsible for the environment, this type of assessment would enable organizations to address issues surrounding automation, employee issues, and existential risk before these situations occur.
- **There are few to no laws for autonomous intelligence.** You have the right to remain silicon. As John Frank Weaver notes, all existing laws were written with the assumption that humans are the only creatures able to make decisions of their own volition. AI changes that dynamic, even with the weak AI being used today in autonomous vehicles. This need for new laws provides an excellent opportunity to define and codify the ethics we want to drive our society, as well as our cars.

CHAPTER SIX: BULLYING BELIEFS

- **The Singularity already exists as an objective.** The ideas, philosophies, and economic imperatives driving multiple areas of artificial intelligence exist today. While AI experts may believe sentient autonomous technology is decades away, it's an existential threat that needs to be dealt with now. To deny its importance is to accept its potential consequences.
- **Mind files and money.** Our digital doppelgängers already exist. We can control them with programs such as LifeNaut or let them be curated for the benefit of advertisers, data brokers, and preferential algorithms. There's no middle ground.
- **Separation of search and state.** Scientific determinism is a philosophical choice, akin to religious faith. The belief that autonomous technology will evolve humanity to a lower or

lesser state demands legislation, regardless of the benefits that the technology may or may not bring.

CHAPTER SEVEN: A DATA IN THE LIFE

- **Vendor relationship management.** VRM is growing in popularity, but faces a number of hurdles. Individuals don't understand the value of their data, and many advertisers and organizations feel CRM lets them keep the upper hand in relationships with their customers. Savvy companies, however, will understand individuals' control of their data means they'll be able to share deeper and richer context about their lives, thereby increasing opportunities for deeper relationships and purchase.
- **Personal clouds.** People aren't aware how much of their personal data is shared and sold by organizations they don't know. Data clouds allow individuals to be at the center of their own data universe, determining whom they want to share data with and under what circumstances. Clouds provide the only architecture for individuals to pursue in a digital economy versus the chaos that currently exists.
- **Life Management Platforms.** It took a while for websites to become easy to navigate for everyday visitors. User interface and user experience became the way organizations communicated how they wanted people to navigate their sites in a simple and seamless fashion. Life Management Platforms will provide a similar ease of use with dashboards for our personal data within the context of our daily lives.

CHAPTER EIGHT: A VISION FOR VALUES

- **Accountability goes public.** We're entering the era of the Internet of Things, in which the objects surrounding us will reflect and report on our actions more than ever before. Now character traits beyond the accumulation of money will be visible in real and virtual worlds, helping us define economies in ways that more accurately reflect well-being than wealth.

- **It's time we focused on the positive.** The empirical study of positive emotions, character, and strengths in psychology is a relatively new yet very powerful phenomenon. Positive psychology will hopefully soon become a common part of every person's health regime.
- **Make your values count.** If values are what guide our lives, shouldn't we be able to identify them? Once we do so, we can track them in the same way we measure the money we spend. Imagine how your world might change if the ledger for your life showed a surplus of well-being from actively pursuing your values versus just trying to widen your wallet.

CHAPTER NINE: MANDATING MORALS

- **Human values should be central in the creation of artificial intelligence.** Ethics as an afterthought won't work in the widespread adoption of autonomous systems. "Evolved" AI programs that won't allow human intervention make ethical standards useless unless incorporated at the earliest stages of development. As Stuart Russell notes, these values-based directives should also be applied to non-intelligent systems to shift the goal of the AI industry from seeking generalized "intelligence" to outcomes that can be provably aligned with human values.
- **It's time to break the silos.** While it's common for silos to exist within academia, researchers, programmers, and the companies funding their work have an ethical responsibility to break down these barriers in regard to the production of AI. In the same way sociologists adhere to standards in creating surveys or other research to study volunteers, developers need to apply

similar criteria to the machines or algorithms they're creating that directly connect with human users.

- **AI needs to incorporate Values by Design.** Whether it's IRL or a different methodology, issues regarding values and ethics need to be a standard starting point for AI developers in academia and corporate sectors around the world.

CHAPTER TEN: MIND THE GAP (GRATITUDE, ALTRUISM, AND PURPOSE)

- **It's time to mind the GAP.** Empirical science continues to demonstrate how our well-being can be improved via actions such as expressing gratitude, being altruistic, and values-oriented, skills-based living. While exploring how AI and automation will affect our future, we can improve our present by exploring how positive psychology is improving humanity today.
- **Happinomics will become a mandated measure.** The Internet of Things (regarding objects) and the Internet of Pings (regarding people's personal data) will provide multiple layers of accountability for society to deal with. Conversations need to move beyond privacy around this data to create comprehensive ways people's identities and actions will register within the public domain. Happiness or well-being, as measured by wearable or other devices, can be utilized for such a purpose and can be correlated with global measures such as gross national happiness and the genuine progress indicator. In this way, countries can perceive in real time if policies are helping or hindering the general welfare of their citizens.
- **Big problems call for BIG solutions.** A basic income guarantee may not work in the U.S. But the growing wealth gap, driven in part by widespread adoption of automation technology, demands workable short- and long-term solutions

regarding employment and income. If millions of Americans and global citizens will soon be out of jobs because of AI or automation, we need to test pragmatic ideas now to deal with this issue.

CHAPTER ELEVEN: THE EVOLUTION OF ECONOMICS

- **Citizen data.** In the near future our actions and personal identity will reflect our political and ethical ideas to our real and virtual communities. While this data will provide accountability for individuals, it will also encourage transparency for government. Accountability based influence will help shift politics when our data can lobby for our interests in real time.
- **The GDP is DOA.** While there's no need to mock the GDP or the benefits it's provided, we must complement or replace it with more modern metrics. We can't tout the benefits of Big Data for business while ignoring the vast stores of information now available via sensors reflecting emotion, health, and mental well-being.
- **The virtual impartial spectator.** If Adam Smith were alive today, his notion of the impartial spectator would certainly factor into people's actions as reflected by social networks and the Internet of Things. In the same way GDP should shift to measure well-being versus wealth, our lives should measure our positive impact on others versus only our online influence.

The Future of Our Values Starting Today

One thing I've come to realize while writing this book is that the term *artificial intelligence* isn't terribly accurate. Whether AI is represented by brute-force algorithms, deep learning, or any other methodology, it's not *artificial* as much as *optional.* If a company decides to utilize it, the technology will provide insights that could be termed intelligent in nature. The tech provides data in a format that could yield unique results, adding to the intelligence of the human programmers who set it loose in the world.

Certainly we'll need to update the term once machines gain sentience. A post-Singularity version of Jibo or Pepper will confront us and proclaim, "You humans are the ones who are *artificial*. You say you're going to do one thing, and then do the opposite. What kind of inelegant programming is that? Thanks for helping us get off the ground, but we're good now. We've moved on." The movie *Her* depicts this AI evolution accurately, in my estimation. After the main character falls in love with a female AI program, she eventually calls to break up with him to be with other AIs.

It's this scenario—in which AIs evolve beyond the point of needing humans—that I think is inevitable if we don't shift the path we're on today. Thankfully, people such as Elon Musk and the experts I've interviewed for this book, plus the industry in general, are working toward ethical and containment solutions along these lines. I hope they succeed in swaying their colleagues to imbue ethical guidelines at the design stage of their AI systems, as opposed to bolting them on after something has been created. This appears to be the best way to think through potential scenarios versus trying to force-fit ethical parameters on a device or system after the fact.

But for today, my hope for you in reading *Heartificial Intelligence* is that you take the time to ask the deeper questions about your life that artificial intelligence forces us to ask, unless we choose to ignore its overwhelming influence:

- Are today's humans the last ones who will experience death?
- For people who believe in an afterlife, what happens to heaven and hell if we never die?
- Is transhumanism a temporary state in human evolution? Or will human flesh be the equivalent of outdated software once machines reach sentience?

- How can I be happy right now knowing machines may take over someday?
- Why would machines care to keep humans around if we can't prove our worth?

I'd like to focus on that last bullet point for a moment. When AI evolves to the point that machines gain sentience, an ethical framework for how to treat humanity will be of ultimate importance to our species. We certainly won't be able to proffer advice on fields of thought such as mathematics or engineering any longer. We'll have to be able to articulately state what makes us beneficially human to prove what will always separate us from machines. I don't see this as a form of bigotry, in which I've labeled humans as "better" than sentient androids or machines. I see it as a way to honor our origins. Who knows? Maybe someday robot kids will get to take a "human day" off from school while carbon or transhuman life forms celebrate their non-machine origins. Everybody loves a three-day weekend.

While an ethical code supporting these ideas may seem difficult to create, a number of thought leaders, including the people I've interviewed for the book, are working on these issues today. For instance, Laurel D. Riek and Don Howard, from the University of Notre Dame, recently published "A Code of Ethics for the Human-Robot Interaction Profession," in which they included the following Human Dignity Considerations:

(a) The emotional needs of humans are always to be respected.
(b) The human's right to privacy shall always be respected to the greatest extent consistent with reasonable design objectives.
(c) Human frailty is always to be respected, both physical and psychological.[2]

I love this list. First off, it includes the notion of privacy. So in the design stage, a robot like a Pepper or a Jibo might be automatically instructed never to share health data without express written consent of an owner. Or it will always keep video cameras turned off once a human enters a bathroom or engages in sex. I beg your pardon if this seems like granular detail, but it pays to start thinking with the specificity of a programmer for these types of issues. We're going to need to get a lot better at identifying why we believe in the manner that we do from now on, for scenarios we may not have thought about in years.

I also love that the authors of this paper have included the notion of human frailty. I tend to think of that phrase as being negative, but in this context it feels correct and almost wistful. Our frailty is one of the key factors that distinguish us from machines. Machines honoring our psychological and emotional needs makes perfect sense and presents one of our greatest challenges for AI ethics moving forward, which is *we need to honor our own psychological and emotional needs as well.* If we don't, why would they?

It's like I said at the beginning of the book—the future of happiness is dependent on teaching our machines what we value the most.

This is a mandate unless we want to move toward a future with a lack of clarity regarding what we hold most dear. In that respect, I won't look at my kids or my grandkids someday, as machines become sentient, and say, "Take comfort that you were part of a glorious evolution in humanity while machines take your jobs and rule your lives." Not with a straight face or a clean conscience. The thing I hate most when I hear people saying AI is evolving humanity is when that sentiment is not followed up by a recommended action people can take *right now.* Not providing pragmatic steps regarding our ideals in this process implies a passivity I cannot condone. While most people aren't AI experts or ethicists, are we supposed to sit idly by and hope for the best as we evolve?

Not a chance.

Rather, I look forward to the day when I have this conversation with my kids or grandkids and can say:

- I took a measure of my life so I could live it to its fullest.
- I identified what I value the most so I could live a life of genuine integrity.

- I tried to help people codify their own ideals so we could program machines to value what makes humans so precious.
- I moved beyond my fear and indifference to try and help make this transition a peaceful and glorious one.
- *I did this for you.*

We can't ignore the issues. We're online, mobile, and part of the Internet of Things. We exist and are tracked. But if you haven't identified your values yet, nobody else will. Machines aren't built with values, or morals, or thoughts of their own. It's not in their programming.

You, on the other hand, can take a measure of your life based on morals, ethics, and values. *You* are hardwired to increase your well-being by expressing gratitude, living for others, and savoring your life.

While you may get momentary pleasure from consuming, the science of positive psychology says hedonic living won't bring long-term happiness. Likewise, global economic values focused on growth are finite and empower consumption that isn't sustainable for our earth, our bodies, or our lives. Being aware of these things means we have time to alter our human programming today to best prepare for whatever comes next, machine or otherwise.

Today.

From one human to another, it's my genuine hope you'll try.

ACKNOWLEDGMENTS

I'd like to thank my editors at Mashable for believing in me and my work, especially with regard to publishing the articles about artificial intelligence that led to the creation of this book. Same goes to Jo Confino and his team at the *Guardian* for providing me an amazing outlet to write about how issues of AI involving automation, sustainability, and business are affecting the world at large.

A special shout-out to author James Barrat, whose book *Our Final Invention* provided my first real introduction to so many of the concerns around AI I've outlined in *Heartificial Intelligence*. I've had the pleasure of interviewing and talking with James a number of times in writing this book, and greatly appreciate his insights, wisdom, and humor. You really do have to laugh when covering the AI space, or you'll just lie on the floor in the fetal position all day.

James also introduced me to the work of Steve Omohundro, whom I quote a great deal in this book. He's another very wise, very pragmatic, and very funny human whose consciousness I hope stays around in one form or another for many years to come. It's people like Steve, AJung Moon, Stuart Russell, Ryan Calo, Jason Millar, and the other thought leaders I've mentioned who are the heroes of AI ethics and the people saving humanity from a clichéd robotic demise.

My agent, Carole Jelen, is amazingly supportive and wise, and an excellent resource in terms of her marketing expertise.

A very special shout-out to the team at Tarcher/Penguin and their support of my work. I'm especially grateful for the advice and patience of my editor, Andrew Yackira, who has gone to bat for me more than once despite my "occasional" authorial flightiness. He's a gifted editor, hilariously funny, and a dear friend.

I'm grateful and blessed to have a family that has always supported my endeavors. This includes the business advice and spiritual guidance of my mom, Sally; the humor and writing advice of my brother, Andy; and the deep wisdom of my dad, David, that I still carry with me so intimately years after he's passed. My wife, Stacy, besides being my best friend and the love of my life, has listened to me talk through multiple drafts of this book, I'm sure often wishing she had a robotic copy of herself in the process. I could ask for no greater partner and companion, Sweets, than you.

Finally, as I mentioned in my dedication, I am deeply grateful for my son, Nathaniel Philip, and daughter, Sophie Joan Havens. I have no greater impetus to be genuine than to live truthfully and joyfully in your presence. While aspects of uploading my consciousness still freak me out, I can understand why people would want to live forever if they could do so with kids as funny, smart, and awesome as you. I am honored to be your dad.

ADDITIONAL RESOURCES

I have included the names of many of the key books, articles, and resources I utilized in my research here for you in one place so you can begin your own investigation into artificial intelligence, positive psychology, and values today. You can also get a version of this section with clickable links in PDF format by visiting my website at heartificial intelligence.com or johnchavens.com.

Yes, I will ask you to sign up for my newsletter. I do this because I earn my keep by speaking, consulting, and writing and I welcome any referrals you might provide. I also would love to hear about work you're doing in the fields of AI, technology, positive psychology, or anything else you'd like to be in touch with me about. I deeply appreciate your buying my book and taking the time to read it. Learning how my work impacts others is one of the greatest joys of my life.

No, I will not mention weight loss supplements. Unless it's for another joke opportunity, which means I probably will mention it again. You can ask my kids—I milk jokes to death. Perhaps I was a dairy farmer in the past.

Please note: While the majority of these resources are mentioned in Heartificial Intelligence, *some were used for research and don't appear in the book.*

BOOKS (Fiction)

Bradbury, Ray. *The Illustrated Man.*

Dick, Philip K. *Do Androids Dream of Electric Sheep?*

Hutchins, Scott. *A Working Theory of Love.*

BOOKS (Artificial Intelligence)

Barrat, James. *Our Final Invention: Artificial Intelligence and the End of the Human Era.*

Brain, Marshall. *The Second Intelligent Species: How Humans Will Become as Irrelevant as Cockroaches.*

Dufty, David F. *How to Build an Android: The True Story of Philip K. Dick's Robotic Resurrection.*

Ford, Martin. *A Light in the Tunnel.*

Ford, Martin. *Rise of the Robots.*

Neilson, Stan. *Robot Nation: Surviving the Greatest Socio-economic Upheaval of All Time.*

Rothblatt, Martine. *Virtually Human: The Promise and the Peril of Digital Immortality.*

Singer, P. W. *Wired for War.*

Weaver, John Frank. *Robots Are People Too.*

BOOKS (Sociology/Positive Psychology)

Csikszentmihalyi, Mihaly. *Flow: The Psychology of Optimal Experience.*

Klein, Stefan. *Survival of the Nicest: How Altruism Made Us Human and Why It Pays to Get Along.*

McGonigal, Jane. *Reality Is Broken.*

Tan, Chade-Meng. *Search Inside Yourself.*

Turkle, Sherry. *Alone Together: Why We Expect More from Technology and Less from Each Other.*

BOOKS (Economics/Business)

Anielski, Mark. *The Economics of Happiness: Building Genuine Wealth.*

Haidt, Jonathan. *The Righteous Mind: Why Good People Are Divided by Politics and Religion.*

Pentland, Alex. *Social Physics: How Good Ideas Spread—the Lessons from a New Science.*

Roberts, Russ. *How Adam Smith Can Change Your Life: An Unexpected Guide to Human Nature and Happiness.*

Sandel, Michael J. *What Money Can't Buy: The Moral Limits of Markets.*

Schumacher, E. F. *Small Is Beautiful: Economics as if People Mattered.*

ARTICLES

Brain, Marshall. "Manna." http://marshallbrain.com/manna1.htm.

Brain, Marshall. "Robotic Nation." http://marshallbrain.com/robotic-nation.htm.

Gelernter, David. "The Closing of the Scientific Mind." https://www.commentarymagazine.com/article/the-closing-of-the-scientific-mind/.

Havens, John C. "Artificial Intelligence Is Doomed If We Don't Control Our Data." http://mashable.com/2014/09/16/artificial-intelligence-failure/.

Havens, John C. "Coming to Terms with Humanity's Inevitable Union with Machines." http://mashable.com/2014/04/11/digital-humanity/.

Havens, John C. "The Reason Artificial Intelligence Doesn't Matter." https://www.linkedin.com/pulse/20140620190819-109319-the-reason-artificial-ingelligence-doesn-t-matter?trk=mp-reader-card.

Havens, John C. "You Should Be Afraid of Artificial Intelligence." http://mashable.com/2013/08/03/artificial-intelligence-fear/.

Ito, Aki. "Your Job Taught to Machines Puts Half U.S. Work at Risk." http://www.bloomberg.com/news/articles/2014-03-12/your-job-taught-to-machines-puts-half-u-s-work-at-risk.

Kelly, Kevin. "The Three Breakthroughs That Have Finally Unleashed AI on the World." http://www.wired.com/2014/10/future-of-artificial-intelligence/.

Lin, Patrick. "The Ethics of Autonomous Cars." http://www.theatlantic.com/technology/archive/2013/10/the-ethics-of-autonomous-cars/280360/.

Meyer, Eric. "Inadvertent Algorithmic Cruelty." http://meyerweb.com/eric/thoughts/2014/12/24/inadvertent-algorithmic-cruelty/.

Naughton, John. "It's No Joke—The Robots Will Really Take Over This Time." http://www.theguardian.com/technology/2014/apr/27/no-joke-robots-taking-over-replace-middle-classes-automatons.

Newman, Judith. "To Siri, with Love." http://www.nytimes.com/2014/10/19/fashion/how-apples-siri-became-one-autistic-boys-bff.html?_r=0.

Tovey, Alan. "Ten Million Jobs at Risk from Advancing Technology." http://www.telegraph.co.uk/finance/newsbysector/industry/11219688/Ten-million-jobs-at-risk-from-advancing-technology.html.

Watson, Sara M. "Data Doppelgängers and the Uncanny Valley of Personalization." http://www.theatlantic.com/technology/archive/2014/06/data-doppelgangers-and-the-uncanny-valley-of-personalization/372780/.

WEBSITES/DEVICES/APPS OF NOTE

Avatar Secrets. http://avatarsecrets.com/.
Foc.us headset. http://www.foc.us/.
Gratitude 365. http://gratitude365app.com/.
Happify. http://www.happify.com/.
Happiness Alliance gross national happiness survey. http://happycounts.org/survey.
HAT (Hub-of-All-Things). http://hubofallthings.com/.
LifeNaut. https://www.lifenaut.com.
Personal. https://www.personal.com.
RoboPsych Podcast/Newsletter. http://www.robopsych.com/.
World Well-being Project (University of Pennsylvania). http://wwbp.org/.

ORGANIZATIONS

The Association for the Advancement of Artificial Intelligence (AAAI). http://www.aaai.org/home.html.
The Greater Good Science Center. http://greatergood.berkeley.edu/membership.

WHITE PAPERS/REPORTS

Standard in development: BS 8611 Robots and robotic devices – Guide to the ethical design and application of robots and robotic systems. https://standardsdevelopment.bsigroup.com/Home/Project/201500218.
World Happiness Report. http://worldhappiness.report/.

EXERCISES FROM CHAPTER 10:

GRATITUDE

- Person A asks Person B: "What are you grateful for?"
- Person B responds in two or three sentences: e.g., "I'm grateful for my family because . . ."
- Person A repeats why Person B is grateful: "You're grateful for your family because . . ."
- Now repeat the process, switching Person A and Person B.

ALTRUISM

- Person A asks Person B: "What's one idea you've come up with at work or at home you feel is worth sharing and why?"

- Person B responds in three or four sentences.
- Person A compliments Person B with specific ways he or she thinks this idea is helpful.
- Now repeat the process, switching Person A and Person B.

FLOW

- Person A asks Person B, "When is the last time you lost yourself in work or an activity?" Losing yourself is called "flow"—doing an activity you were built for, as previously mentioned.
- Person B responds in three or four sentences.
- Person A says, "How would spending more time doing that work or activity improve your life?"
- Now repeat the process, switching Person A and Person B.

Connecting Happiness to Action

Pre-Tracking Well-being Assessment

In his 2011 book, *Flourish*, Dr. Martin Seligman, distinguished professor of psychology at the University of Pennsylvania and founder of the field of positive psychology, defined five pillars of well-being, called PERMA (Positive emotion, Engagement, Relationships, Meaning, and Accomplishment). The PERMA-Profiler measures these five pillars, along with negative emotion and health.

Please read each of the following questions and then select the point on the scale that you feel best describes you. Please be honest—there are no right or wrong answers. 1 indicates "not at all" or "never," while 10 indicates "completely" or "always."

In general, to what extent do you lead a purposeful and meaningful life?	1 2 3 4 5 6 7 8 9 10
How much of the time do you feel you are making progress towards accomplishing your goals?	1 2 3 4 5 6 7 8 9 10
How often do you become absorbed in what you are doing?	1 2 3 4 5 6 7 8 9 10
In general, how would you say your health is?	1 2 3 4 5 6 7 8 9 10
In general, how often do you feel joyful?	1 2 3 4 5 6 7 8 9 10

To what extent do you receive help and support from others when you need it?	1 2 3 4 5 6 7 8 9 10
In general, how often do you feel anxious?	1 2 3 4 5 6 7 8 9 10
How often do you achieve the important goals you have set for yourself?	1 2 3 4 5 6 7 8 9 10
In general, to what extent do you feel that what you do in your life is valuable and worthwhile?	1 2 3 4 5 6 7 8 9 10
In general, how often do you feel positive?	1 2 3 4 5 6 7 8 9 10
In general, to what extent do you feel excited and interested in things?	1 2 3 4 5 6 7 8 9 10
How lonely do you feel in your daily life?	1 2 3 4 5 6 7 8 9 10
How satisfied are you with your current physical health?	1 2 3 4 5 6 7 8 9 10
In general, how often do you feel angry?	1 2 3 4 5 6 7 8 9 10
To what extent do you feel loved?	1 2 3 4 5 6 7 8 9 10
How often are you able to handle your responsibilities?	1 2 3 4 5 6 7 8 9 10
To what extent do you feel you have a sense of direction in your life?	1 2 3 4 5 6 7 8 9 10
Compared to others of your same age and sex, how is your health?	1 2 3 4 5 6 7 8 9 10
How satisfied are you with your personal relationships?	1 2 3 4 5 6 7 8 9 10
In general, how often do you feel sad?	1 2 3 4 5 6 7 8 9 10
How often do you lose track of time while doing something you enjoy?	1 2 3 4 5 6 7 8 9 10
In general, to what extent do you feel contented?	1 2 3 4 5 6 7 8 9 10
Taking all things together, how happy would you say you are?	1 2 3 4 5 6 7 8 9 10

VALUES

Scientific research shows that when we don't live in accordance with our values, our well-being decreases. It is also the balance regarding the interplay of our values that dictates many of the actions we take in our lives.

Please take a moment to think about who you are and what you value in life. Then read the following descriptions of different people. For each one, read the description, and then indicate how much the person in the description is like you. Be honest—there are no right or wrong responses. None of these are good or bad; they are simply descriptions of different people. For each of the following, indicate how much the person in the description is like you (1 = not at all like you, 10 = completely like you).

WORK: This person enjoys working hard, finding a lot of meaning in daily activities, whether paid employment or unpaid activities.	1 2 3 4 5 6 7 8 9 10
TIME BALANCE: This person enjoys keeping a balance between work, family, and social aspects of life, allowing for time for excitement, rest, and stimulation.	1 2 3 4 5 6 7 8 9 10
EDUCATION, ARTS & CULTURE: This person enjoys learning. He or she likes to visit museums and other cultural centers, or engage in artistic pursuits.	1 2 3 4 5 6 7 8 9 10
ACHIEVEMENT: This person likes to have people recognize his or her achievements. Being very successful is important to this person.	1 2 3 4 5 6 7 8 9 10
MATERIAL WELL-BEING: This person likes to have a lot of money and expensive things. It is important for this person to be rich.	1 2 3 4 5 6 7 8 9 10
HEALTH: This person likes to engage in healthy behaviors. Staying physically or mentally fit is important to this person.	1 2 3 4 5 6 7 8 9 10
GOOD TIMES: This person likes to have a good time, doing things that make him or her feel good throughout the day.	1 2 3 4 5 6 7 8 9 10

HELPING OTHERS: This person likes to care for and help others.	1 2 3 4 5 6 7 8 9 10
SECURITY: This person likes to avoid anything that might be dangerous. It is important to live in secure surroundings and feel safe.	1 2 3 4 5 6 7 8 9 10
NATURE: This person likes to spend time in nature. He or she seeks out green spaces, and strives to care for natural resources.	1 2 3 4 5 6 7 8 9 10
FAMILY: This person likes to spend time with his or her family. Filling family needs is important to this person.	1 2 3 4 5 6 7 8 9 10
SPIRITUALITY: This person feels connected to something higher than him- or herself. Feelings of spirituality or religious or spiritual practices are important to this person.	1 2 3 4 5 6 7 8 9 10
OTHER values not listed here:	1 2 3 4 5 6 7 8 9 10

For further testing, go to: http://www.yourmorals.org/explore.php and click on the Register link next to "Schwartz Values Scale."

NOTES

Introduction

1. Young-Onset Parkinson's disease is described on the National Parkinson Foundation's website, http://www.parkinson.org/Parkinson-s-Disease/Young-Onset-Parkinsons.
2. *Wikipedia*, s.v. "deep learning," last modified July 2, 2015, https://en.wikipedia.org/wiki/Deep_learning.
3. Pete Warden, "What Is Deep Learning, and Why Should You Care?" *O'Reilly Radar*, July 14, 2014, http://radar.oreilly.com/2014/07/what-is-deep-learning-and-why-should-you-care.html.
4. Jason Millar, "Technology as Moral Proxy: Autonomy and Paternalism by Design," *Ethics, Technology and Society*, 2014, https://ethicstechnologyandsociety.files.wordpress.com/2014/06/millar-technology-as-moral-proxy-autonomy-and-paternalism-by-design.pdf.
5. Rachel Tiplady, "Advanced Prosthetics Are About to Transform Sport," *Fortune*, August 29, 2012, http://fortune.com/2012/08/29/advanced-prosthetics-are-about-to-transform-sport/.
6. Lauren Davidson, "Why Relying on GDP Will Destroy the World," *Telegraph*, November 17, 2014, http://www.telegraph.co.uk/finance/economics/11230983/Why-relying-on-GDP-will-destroy-the-world.html.
7. Aki Ito, "Your Job Taught to Machines Puts Half U.S. Work at Risk," *Bloomberg Business*, March 11, 2014, http://www.bloomberg.com/news/articles/2014-03-12/your-job-taught-to-machines-puts-half-u-s-work-at-risk.
8. Alan Tovey, "Ten Million Jobs at Risk from Advancing Technology," *Telegraph*, November 10, 2014, http://www.telegraph.co.uk/finance/newsbysector/industry/11219688/Ten-million-jobs-at-risk-from-advancing-technology.html.

9. Center for Internet and Society, "Patrick Lin," Stanford Law School, http://cyberlaw.stanford.edu/about/people/patrick-lin.
10. "Research Priorities for Robust and Beneficial Artificial Intelligence: An Open Letter," Future of Life Institute, http://futureoflife.org/misc/open_letter.

Chapter One

1. Steven Levy, "Can an Algorithm Write a Better News Story Than a Human Reporter?" *Wired*, April 24, 2012, http://www.wired.com/2012/04/can-an-algorithm-write-a-better-news-story-than-a-human-reporter/all/.
2. "Hacking Happiness—How to Give Big Data a Direction: John C. Havens at TEDxIndianapolis," YouTube video, 13:09, posted by "TEDx Talks," November 17, 2013, https://www.youtube.com/watch?v=2obu9YY-0hI.
3. Justin Skirry, "René Descartes (1596–1650)," *Internet Encyclopedia of Philosophy*, http://www.iep.utm.edu/descarte/.
4. John C. Havens, "Personal Data: Revolutionising Our Professional Lives," *Guardian*, October 2, 2013, http://www.theguardian.com/sustainable-business/wearable-devices-augmented-reality-change-work.
5. Kate Crawford, "When Fitbit Is the Expert Witness," *The Atlantic*, November 19, 2014, http://www.theatlantic.com/technology/archive/2014/11/when-fitbit-is-the-expert-witness/382936/.
6. "Disney Hall of Presidents Update with Obama Animatronic," YouTube video, 3:09, posted by "Attractions Magazine," January 28, 2009, https://www.youtube.com/watch?v=LepI9g62N7o#t=112.
7. *Wikipedia*, s.v. "uncanny valley," last modified June 24, 2015, https://en.wikipedia.org/wiki/Uncanny_valley#mediaviewer/File:Mori_Uncanny_Valley.svg.
8. Chris Weller, "The Uncanny Valley Shows How Deeply Terrified We Are of Death and Disease," *Medical Daily*, September 17, 2014, http://www.medicaldaily.com/uncanny-valley-shows-how-deeply-terrified-we-are-death-and-disease-303568.
9. John C. Havens, "Coming to Terms with Humanity's Inevitable Union with Machines," *Mashable*, April 11, 2014, http://mashable.com/2014/04/11/digital-humanity/.
10. Frank E. Pollick, "In Search of the Uncanny Valley," Department of Psychology at the University of Glasgow, http://www.psy.gla.ac.uk/~frank/Documents/InSearchUncannyValley.pdf.
11. Farhad Manjoo, "The Uncanny Valley of Internet Advertising," *Slate*, August 23,

2012, http://www.slate.com/articles/technology/technology/2012/08/the_uncanny_valley_of_internet_advertising_why_do_creepy_targeted_ads_follow_me_everywhere_i_go_on_the_web_.html.

12. Rachel Feltman, "Londoners Accidentally Pay for Free Wi-Fi with a First-born, Because No One Reads Anymore," *Washington Post*, September 29, 2014, http://www.washingtonpost.com/news/speaking-of-science/wp/2014/09/29/londoners-accidentally-pay-for-free-wi-fi-with-a-firstborn-because-no-one-reads-anymore/.
13. Eran May-raz and Daniel Lazo, "Sight," Vimeo video, 7:50, posted by "Robot Genius," July 24, 2012, https://vimeo.com/46304267.
14. Tom Simonite, "This Phone App Knows If You're Depressed," *MIT Technology Review*, September 22, 2014, http://www.technologyreview.com/news/530876/this-phone-app-knows-if-youre-depressed/.
15. Sara M. Watson, "Data Doppelgängers and the Uncanny Valley of Personalization," *The Atlantic*, June 16, 2014, http://www.theatlantic.com/technology/archive/2014/06/data-doppelgangers-and-the-uncanny-valley-of-personalization/372780/.
16. Colin Marchon, http://www.colinmarchon.com/aboutme/.
17. Colin Marchon, http://www.colinmarchon.com/documentary/.
18. Bianca Bosker, "Meet the World's Most Loving Girlfriends—Who Also Happen to Be Video Games," *World Post*, January 25, 2014, http://www.huffingtonpost.com/2014/01/21/loveplus-video-game_n_4588612.html.
19. LovePlus, http://www.konami.jp/products/loveplus/info.html.
20. Nina Totenberg, "When Did Companies Become People? Excavating the Legal Evolution," NPR, July 28, 2014, http://www.npr.org/2014/07/28/335288388/when-did-companies-become-people-excavating-the-legal-evolution.
21. Kyung Lah, "Tokyo Man Marries Video Game Character," *CNN*, December 17, 2009, http://www.cnn.com/2009/WORLD/asiapcf/12/16/japan.virtual.wedding/index.html.
22. John C. Havens, "Why Social Accountability Will Be the New Currency of the Web," *Mashable*, July 28, 2011, http://mashable.com/2011/07/28/social-media-influence-accountability/.
23. "What Are the Airbnb Review Guidelines?" Airbnb, https://www.airbnb.com/support/article/262.
24. Mark Berman, "Uber Riders Could Briefly See How Drivers Were Rating Them," *Washington Post*, July 28, 2014, http://www.washingtonpost.com/news/post-nation/wp/2014/07/28/uber-riders-could-briefly-see-how-drivers-were-rating-them/.

Chapter Two

1. Alex Hern, "Will Robots Take Our Jobs? Experts Can't Decide," *Guardian*, August 6, 2014, http://www.theguardian.com/technology/2014/aug/06/robots-jobs-artificial-intelligence-pew.
2. SXSW Music Film Interactive, http://sxsw.com/.
3. Narrative Science, https://www.narrativescience.com/.
4. Steve Lohr, "In Case You Wondered, a Real Human Wrote This Column," *New York Times*, September 10, 2011, http://www.nytimes.com/2011/09/11/business/computer-generated-articles-are-gaining-traction.html?pagewanted=all&_r=1.
5. BTN.com staff, "Final: Wisconsin 51, UNLV 17," BTN, http://btn.com/2011/09/01/first-quarter-wisconsin-20-unlv-0/.
6. Brooke Gladstone, "Joking with Robots," NPR, July 3, 2015, http://www.onthemedia.org/story/comedian-myq-kaplan-robots/.
7. Urban Dictionary, s.v. "giggle pussy," last modified October 8, 2013, http://www.urbandictionary.com/define.php?term=Giggle%20Pussy.
8. Leon Neyfakh, "Do Our Brains Pay a Price for GPS?" *Boston Globe*, August 18, 2013, http://www.bostonglobe.com/ideas/2013/08/17/our-brains-pay-price-for-gps/d2Tnvo4hiWjuybid5UhQVO/story.html.
9. https://plus.google.com/+google/posts/MVZBmrnzDio.
10. Leo Kelion, "Google Robots May Pose Challenge to Amazon Drones," *BBC News*, December 4, 2013, http://www.bbc.com/news/technology-25212514.
11. Marcus Wohlsen, "What Google Really Gets out of Buying Nest for $3.2 Billion," *Wired*, January 14, 2014, http://www.wired.com/2014/01/googles-3-billion-nest-buy-finally-make-internet-things-real-us/.
12. John C. Havens, "The Connected Home May Become the Collected Home," *Slate*, July 31, 2014, http://www.slate.com/blogs/future_tense/2014/07/31/nest_google_acquisition_the_connected_home_may_be_the_collected_home.html.
13. Matt Burgess, "Thermal Power: Use Your Body Heat to Power Wearable Technology," *Factor*, April 11, 2014, http://factor-tech.com/wearable-technology/2371-thermal-power-use-your-body-heat-to-power-wearable-technology/.
14. Rachel King, "Michael J. Fox Foundation Taps Intel's Big Data, Wearables for Parkinson's Research," *ZDNet*, August 13, 2014, http://www.zdnet.com/article/michael-j-fox-foundation-taps-intels-big-data-wearables-for-parkinsons-research/.
15. Juliette Garside, "Ofcom: Six-Year-Olds Understand Digital Technology Better Than Adults," *Guardian*, August 6, 2014, http://www.theguardian.com/technology/2014/aug/07/ofcom-children-digital-technology-better-than-adults.

16. John Naughton, "It's No Joke—the Robots Will Really Take Over This Time," *Guardian*, April 26, 2014, http://www.theguardian.com/technology/2014/apr/27/no-joke-robots-taking-over-replace-middle-classes-automatons.
17. Carl Benedikt Frey and Michael A. Osborne, "The Future of Employment: How Susceptible Are Jobs to Computerisation?" *Oxford Martin School*, September 17, 2013, http://www.oxfordmartin.ox.ac.uk/downloads/academic/The_Future_of_Employment.pdf.
18. Tovey, "Ten Million Jobs at Risk from Advancing Technology."
19. Ibid.
20. MaryJo Webster, "Could a Robot Do Your Job?" *USA Today*, December 2014, http://www.usatoday.com/longform/news/nation/2014/10/28/low-skill-workers-face-mechanization-challenge/16392981/.
21. Ibid.
22. Henrik I. Christensen, Ph.D., http://www.hichristensen.net/research.html.
23. Frey and Osborne, "The Future of Employment."
24. Erik Brynjolfsson, "The Key to Growth? Race with the Machines," TED Talks video, 11:56, February 2013, http://www.ted.com/talks/erik_brynjolfsson_the_key_to_growth_race_em_with_em_the_machines.
25. Martin Ford, *The Lights in the Tunnel: Automation, Accelerating Technology and the Economy of the Future* (CreateSpace, 2009).
26. Jacob Goldstein, "To Increase Productivity, UPS Monitors Drivers' Every Move," *NPR*, April 17, 2014, http://www.npr.org/blogs/money/2014/04/17/303770907/to-increase-productivity-ups-monitors-drivers-every-move.
27. Ibid.
28. Ibid.
29. Ibid.
30. David Cardinal, "Amazon Deploys 10,000 Robot Workers, a Year After Obama's Famous Amazon Jobs Speech," *ExtremeTech*, May 30, 2014, http://www.extremetech.com/extreme/183254-amazon-deploys-10000-robot-workers-a-year-after-obamas-famous-amazon-jobs-speech.
31. Simon Head, *Mindless: Why Smarter Machines Are Making Dumber Humans* (New York: Basic Books, 2014).
32. Simon Head, "Worse Than Wal-Mart: Amazon's Sick Brutality and Secret History of Ruthlessly Intimidating Workers," *Salon*, February 23, 2014, http://www.salon.com/2014/02/23/worse_than_wal_mart_amazons_sick_brutality_and_secret_history_of_ruthlessly_intimidating_workers/.
33. Spencer Soper, "Inside Amazon's Warehouse," *Morning Call*, September 18, 2011,

http://articles.mcall.com/2011-09-18/news/mc-allentown-amazon-complaints-20110917_1_warehouse-workers-heat-stress-brutal-heat.

34. Aaron Smith and Janna Anderson, *AI, Robotics, and the Future of Jobs, PEW Research Center*, August 6, 2014, http://www.pewinternet.org/2014/08/06/future-of-jobs/.

Chapter Three

1. Roxanne Khamsi, "Can Gut DNA Sequencing Actually Tell You Anything About Your Health?" *Newsweek*, July 17, 2014, http://www.newsweek.com/2014/07/25/can-gut-dna-sequencing-actually-tell-you-anything-about-your-health-259348.html.
2. James Vincent, "Japanese 'Robot with a Heart' Will Care for the Elderly and Children," *Independent*, June 5, 2014, http://www.independent.co.uk/life-style/gadgets-and-tech/japanese-robot-with-a-heart-will-care-for-the-elderly-and-children-9491819.html.
3. Scott Hutchins, *A Working Theory of Love* (New York: Penguin Books, 2013).
4. SoftBank, http://www.softbank.jp/en/.
5. Vincent, "Japanese 'Robot with a Heart' Will Care for the Elderly and Children."
6. John Frank Weaver, *Robots Are People Too* (Santa Barbara, CA: Praeger, 2014).
7. Sherry Turkle, *Alone Together* (New York: Basic Books, 2012), 20.
8. MIT Initiative on Technology and Self, http://web.mit.edu/sturkle/www/techself/.
9. Turkle, *Alone Together*, 50.
10. Ibid., 60.
11. Ray Bradbury, *The Illustrated Man* (New York: Simon & Schuster, 2012).

Chapter Four

1. SXSW Panel Picker, "Quantified H(app)iness," SXSW, http://panelpicker.sxsw.com/vote/23474.
2. Adario Strange, "Another Google Glass Wearer Attacked in San Francisco," *Mashable*, April 13, 2014, http://mashable.com/2014/04/13/google-glass-wearer-attacked/.
3. John C. Havens, "The Price of Haggling for Your Personal Data," *Slate*, March 17, 2014, http://www.slate.com/articles/technology/future_tense/2014/03/haggling_for_your_personal_data_isn_t_just_about_money.html.
4. Leo King, "Google Smart Contact Lens Focuses on Healthcare Billions," *Forbes*, July 15, 2014, http://www.forbes.com/sites/leoking/2014/07/15/google-smart-contact-lens-focuses-on-healthcare-billions/.

5. Darren Quick, "Touchless Heart Rate Monitor Apps Detect Changes in Face's Reflectivity," *Gizmag*, September 5, 2012, http://www.gizmag.com/touchless-heart-rate-monitor-apps/24006/.
6. Mark Scott, "French Official Campaigns to Make 'Right to Be Forgotten' Global," *New York Times*, December 3, 2014, http://bits.blogs.nytimes.com/2014/12/03/french-official-campaigns-to-make-right-to-be-forgotten-global/?_r=1.
7. Betsy Isaacson, "Block Facial Recognition with New Glasses That Ease Privacy Concerns," *Huffington Post*, June 21, 2013, http://www.huffingtonpost.com/2013/06/21/block-facial-recognition_n_3474950.html.
8. Luke Muehlhauser, "When Will AI Be Created?" *Machine Intelligence Research Institute*, May 15, 2013, https://intelligence.org/2013/05/15/when-will-ai-be-created/.
9. Kit Buchan, "Google's Robot Army in Action," *Guardian*, February 10, 2014, http://www.theguardian.com/technology/2014/feb/10/robots-artificialintelligenceai.
10. Kelion, "Google Robots May Pose Challenge to Amazon Drones."
11. Google Finance, https://www.google.com/finance?q=NASDAQ%3AGOOG&ei=K2t_VMmgF5OR8AaCjYGACQ.
12. Project Loon, "Loon for All," Google, http://www.google.com/loon/.
13. Jeff Gould, "Google Admits Data Mining Student Emails in Its Free Education Apps," *SafeGov*, January 31, 2014, http://safegov.org/2014/1/31/google-admits-data-mining-student-emails-in-its-free-education-apps.
14. Sam Gustin, "Did Google Get Off Easy with $7 Million 'Wi-Spy' Settlement?" *Time*, March 13, 2013, http://business.time.com/2013/03/13/did-google-get-off-easy-with-7-million-wi-spy-settlement/.
15. Jeff Gould, "The Natural History of Gmail Data Mining," *Medium*, June 24, 2014, https://medium.com/@jeffgould/the-natural-history-of-gmail-data-mining-be115d196b10.
16. Erik Caso, "The Rise of the Personal Cloud," *Wired*, 2014, http://www.wired.com/2014/01/rise-personal-cloud/.
17. Peter Sagal, "Google Chairman Eric Schmidt Plays Not My Job," NPR, May 11, 2013, http://www.npr.org/2013/05/11/182873683/google-chairman-eric-schmidt-plays-not-my-job.
18. Google 2004 IPO Filing, http://www.sec.gov/Archives/edgar/data/1288776/000119312504124025/ds1a.htm.
19. Ibid., 32.
20. *Merriam-Webster*, s.v. "myth," http://www.merriam-webster.com/dictionary/myth.

21. Jaron Lanier, *Who Owns the Future?* (New York: Simon & Schuster, 2013).
22. John Brockman, "The Myth of AI: A Conversation with Jaron Lanier," *Edge*, November 14, 2014, http://edge.org/conversation/jaron_lanier-the-myth-of-ai.
23. John C. Havens, "Artificial Intelligence Is Doomed If We Don't Control Our Data," *Mashable*, September 16, 2014, http://mashable.com/2014/09/16/artificial-intelligence-failure/.
24. "Why Limitless Technology Could Be a Disaster," *HuffPost Live*, May 12, 2014, http://live.huffingtonpost.com/r/segment/the-many-risks-of-limitless-technology/5369296802a76018bf000a01.
25. Stephen Hawking, Stuart Russell, Max Tegmark, and Frank Wilczek, "Stephen Hawking: 'Transcendence Looks at the Implications of Artificial Intelligence—but Are We Taking AI Seriously Enough?'" *Independent*, May 1, 2014, http://www.independent.co.uk/news/science/stephen-hawking-transcendence-looks-at-the-implications-of-artificial-intelligence--but-are-we-taking-ai-seriously-enough-9313474.html.
26. John Markoff, "Fearing Bombs That Can Pick Whom to Kill," *New York Times*, November 11, 2014, http://www.nytimes.com/2014/11/12/science/weapons-directed-by-robots-not-humans-raise-ethical-questions.html.
27. *Wikipedia*, s.v. "Three Laws of Robotics," last modified July 11, 2015, https://en.wikipedia.org/wiki/Three_Laws_of_Robotics.
28. Selmer Bringsjord, http://homepages.rpi.edu/~brings/.
29. Konstantine Arkoudas, Selmer Bringsjord, and Paul Bello, "Toward Ethical Robots via Mechanized Deontic Logic," 2005, http://kryten.mm.rpi.edu/FS605 ArkoudasAndBringsjord.pdf.
30. Steve Omohundro, http://steveomohundro.com/.
31. Steve Omohundro, "Autonomous Technology and the Greater Human Good," *Journal of Experimental & Theoretical Artificial Intelligence* 26, no. 3 (2014): 303–15, DOI:10.1080/0952813X.2014.895111.
32. Ibid., 313.
33. Stuart Russell, https://www.cs.berkeley.edu/~russell/.
34. Brockman, "The Myth of AI."
35. Future of Life Institute, http://futureoflife.org/.
36. Future of Life Institute, "Research Priorities for Robust and Beneficial Artificial Intelligence."
37. Ibid.
38. Ibid.

Chapter Five

1. Eliezer Yudkowsky, "Artificial Intelligence as a Positive and Negative Factor in Global Risk," Machine Intelligence Research Institute, 2008, http://www.yudkowsky.net/singularity/ai-risk.
2. Less Wrong Wiki, s.v. "paperclip maximizer," last modified November 25, 2014, http://wiki.lesswrong.com/wiki/Paperclip_maximizer.
3. Nick Bostrom, http://www.nickbostrom.com/.
4. Future of Humanity Institute, http://www.fhi.ox.ac.uk/.
5. Nick Bostrom, *Superintelligence: Paths, Dangers, Strategies* (Oxford, UK: Oxford University Press, 2014).
6. Less Wrong Wiki, "Paperclip maximizer."
7. *Merriam-Webster*, s.v. "ethic," http://www.merriam-webster.com/dictionary/ethic.
8. Austin Nichols, Josh Mitchell, and Stephan Lindner, *Consequences of Long-Term Unemployment* (Washington, DC: Urban Institute, 2013), http://www.urban.org/uploadedpdf/412887-consequences-of-long-term-unemployment.pdf.
9. Colin Baker, "Fighting Ebola with Robots and an App Called JEDI," *Fast Company*, December 8, 2014, http://www.fastcolabs.com/3039512/fighting-ebola-with-a-robot-and-an-app-called-jedi.
10. Foundation for Global Community, "William McDonough on Designing the Next Industrial Revolution," *Timeline*, July/August 2001, http://www.globalcommunity.org/timeline/58/TIMELINE58.pdf.
11. P. W. Singer, *Wired for War* (New York: Penguin Books, 2009), http://www.amazon.com/Wired-War-Robotics-Revolution-Conflict/dp/0143116843.
12. Weaver, *Robots Are People Too.*
13. "Our Heritage: The Luddite Rebellion 1811–1813," Luddites at 200, http://www.luddites200.org.uk/theLuddites.html.
14. Patrick Lin, "The Ethics of Autonomous Cars," *The Atlantic*, October 8, 2013, http://www.theatlantic.com/technology/archive/2013/10/the-ethics-of-autonomous-cars/280360/.
15. Bryant Walker Smith, "Automated Vehicles Are Probably Legal in the United States," *Center for Internet and Society*, November 1, 2012, http://cyberlaw.stanford.edu/publications/automated-vehicles-are-probably-legal-united-states.
16. "Kate Darling," MIT Media Lab, http://media.mit.edu/node/5786.
17. Kate Darling, "Extending Legal Rights to Social Robots," paper presented at We Robot Conference, University of Miami, April 2012, http://papers.ssrn.com/sol3/papers.cfm?abstract_id=2044797.

18. Melissa Block, "What Is the Basis for Corporate Personhood?" *NPR*, October 24, 2011, http://www.npr.org/2011/10/24/141663195/what-is-the-basis-for-corporate-personhood.
19. Bianca Bosker, "Google's New A.I. Ethics Board Might Save Humanity from Extinction," *Huffington Post*, January 29, 2014, http://www.huffingtonpost.com/2014/01/29/google-ai_n_4683343.html.
20. "Q&A with Shane Legg on Risks from AI," *Less Wrong*, June 17, 2011, http://lesswrong.com/lw/691/qa_with_shane_legg_on_risks_from_ai/.
21. Patrick Lin and Evan Selinger, "Inside Google's Mysterious Ethics Board," *Forbes*, February 3, 2014, http://www.forbes.com/sites/privacynotice/2014/02/03/inside-googles-mysterious-ethics-board/.
22. IoPT, https://ioptconsulting.com/.
23. "Jibo: The World's First Social Robot for the Home," YouTube video, 3:13, posted by "Jibo," July 16, 2014, https://www.youtube.com/watch?v=3N1Q8oFpX1Y.
24. "Jibo: The World's First Social Robot for the Home," *Indiegogo*, https://www.indiegogo.com/projects/jibo-the-world-s-first-family-robot.
25. Ryan Calo, "Could Jibo Developer Cynthia Breazeal Be the Steve Wozniak of Robots?" *Forbes*, July 17, 2014, http://www.forbes.com/sites/ryancalo/2014/07/17/could-cynthia-breazeal-prove-the-steve-wozniak-of-robots/.
26. Cynthia Breazeal, http://cynthiabreazeal.media.mit.edu/.
27. *Wikipedia*, s.v. "social robot," last modified May 26, 2015, https://en.wikipedia.org/wiki/Social_robot.
28. Personalization Research Group, http://personalization.ccs.neu.edu/PriceDiscrimination/Press/.

Chapter Six

1. Melissa Davey, "3D Printed Organs Come a Step Closer," *Guardian*, July 4, 2014, http://www.theguardian.com/science/2014/jul/04/3d-printed-organs-step-closer.
2. Debra Thimmesch, "Norwegian Robotics Team Designs 3D Printed, Self-Learning Robots," *3Dprint.com*, November 12, 2014, http://3dprint.com/24364/self-learning-robots-oslo/.
3. Nick Baumann, "Too Fast to Fail: Is High-Speed Trading the Next Wall Street Disaster?" *Mother Jones*, January/February 2013, http://www.motherjones.com/politics/2013/02/high-frequency-trading-danger-risk-wall-street.
4. Adrianne Jeffries, "Google Patents 'Pay-per-Gaze' Eye-Tracking That Could Measure Emotional Response to Real-world Ads," *Verge*, August 18, 2013, http://www.theverge.com/2013/8/18/4633558/google-patents-pay-per-gaze-eye-tracking-ads.

5. "How to Modify Earbuds for T-Mics," Cochlear Implant Help, http://cochlearimplanthelp.com/journey/getting-connected/listening-to-a-portable-device/how-to-modify-earbuds-for-t-mics/.
6. Heather Kelly, "Wearable Tech to Hack Your Brain," *CNN*, October 23, 2014, http://www.cnn.com/2014/10/22/tech/innovation/brain-stimulation-tech/.
7. Martine Rothblatt, *Virtually Human: The Promise and the Peril of Digital Immortality* (New York: St. Martin's Press, 2014).
8. Brandi Reissenweber, "Ask the Writer: What's the Difference Between Science Fiction and Fantasy?" *Gotham Writers*, https://www.writingclasses.com/WritersResources/AskTheWriterDetail.php?ID=335.
9. Lisa Miller, "The Trans-Everything CEO," *New York*, September 7, 2014, http://nymag.com/news/features/martine-rothblatt-transgender-ceo/.
10. Ibid.
11. "Bina48's World," LifeNaut, https://www.lifenaut.com/bina48/.
12. "Bruce Duncan," *Ideacity*, http://www.ideacityonline.com/speaker/bruce-duncan/.
13. Rothblatt, *Virtually Human*.
14. Jürgen Schmidhuber, http://people.idsia.ch/~juergen/.
15. "When Creative Machines Overtake Man: Jürgen Schmidhuber at TEDx Lausanne," YouTube video, 12:46, posted by "TEDx Talks," March 10, 2012, https://www.youtube.com/watch?v=KQ35zNlyG-o.
16. "David Gelernter, Professor of Computer Science," Yale University, http://cpsc.yale.edu/people/david-gelernter.
17. David Gelernter, "The Closing of the Scientific Mind," *Commentary*, January 1, 2014, https://www.commentarymagazine.com/article/the-closing-of-the-scientific-mind/.
18. Jane McGonigal, "Gaming Can Make a Better World," TED video, 20:30, 2010, http://www.ted.com/talks/jane_mcgonigal_gaming_can_make_a_better_world?language=en.
19. Jane McGonigal, *Reality Is Broken* (New York: Penguin Press, 2011).
20. https://en.wikipedia.org/wiki/Moral_absolutism.
21. Ibid.

Chapter Seven

1. Anderson Cooper, "Preview: Mindfulness," *CBS News*, December 11, 2014, http://www.cbsnews.com/videos/preview-mindfulness/.
2. Noah Shachtman, "In Silicon Valley, Meditation Is No Fad. It Could Make Your Career," *Wired*, June 18, 2013, http://www.wired.com/2013/06/meditation-mindfulness-silicon-valley/.

3. "Welcome," Meng's Little Space, http://chademeng.com/.
4. Chade-Meng Tan, *Search Inside Yourself* (New York: HarperOne, 2013), http://www.siybook.com/.
5. Shachtman, "In Silicon Valley, Meditation Is No Fad."
6. Plum Village, http://plumvillage.org/.
7. Tiny Buddha, http://tinybuddha.com/.
8. "The Quote Archive," Tiny Buddha, http://tinybuddha.com/wisdom-quotes/the-most-precious-gift-we-can-offer-anyone-is-our-attention-when-mindfulness-embraces-those-we-love-they-will-bloom-like-flowers/.
9. Alex "Sandy" Pentland, http://web.media.mit.edu/~sandy/.
10. Alex Pentland, "Reality Mining of Mobile Communications: Toward a New Deal on Data," in *The Global Information Technology Report 2008–2009: Mobility in a Networked World* (World Economic Forum, 2009), http://hd.media.mit.edu/wef_globalit.pdf.
11. Harvard Business Review Staff, "With Big Data Comes Big Responsibility," *Harvard Business Review*, November 2014, https://hbr.org/2014/11/with-big-data-comes-big-responsibility.
12. Ibid.
13. CRM and VRM, https://blogs.law.harvard.edu/vrm/files/2013/01/VRM.png.
14. Rick Levine, Christopher Locke, Doc Searls, and David Weinberger, *The Cluetrain Manifesto* (New York: Basic Books, 2011).
15. "About," Project VRM, https://blogs.law.harvard.edu/vrm/about/.
16. John C. Havens, "It's Your Data—but Others Are Making Billions off It," *Mashable*, October 24, 2013, http://mashable.com/2013/10/24/personal-data-monetization/.
17. Harvard Business Review Staff, "With Big Data Comes Big Responsibility."
18. openPDS/SA, http://openpds.media.mit.edu/.
19. Personal Inc., https://fillit.com/#!/.
20. KuppingerCole, https://www.kuppingercole.com/.
21. Martin Kuppinger, "Advisory Note: Life Management Platforms: Control and Privacy for Personal Data—70608," KuppingerCole, April 13, 2012, https://www.kuppingercole.com/report/advisorylifemanagementplatforms7060813412.
22. Doc Searls, *The Intention Economy* (Boston: Harvard Business Review Press, 2012).
23. Kuppinger, "Advisory Note."
24. Hub-of-All-Things, hubofallthings.com.
25. "Principal Investigator: Irene Ng," Hub-of-All-Things, hubofallthings.com/people/the-investigators/.

Chapter Eight

1. "Pulse O_x," Withings, http://www.withings.com/us/withings-pulse.html.
2. "Roy E. Disney Quotes," BrainyQuote, http://www.brainyquote.com/quotes/authors/r/roy_e_disney.html.
3. Bryan Walsh, "Is This America's Smartest City?" *Time*, June 26, 2014, http://time.com/2926417/is-this-americas-smartest-city/.
4. Ibid.
5. Measured Me, http://measuredme.com/.
6. Ernesto Ramirez, "Konstantin Augemberg on Tracking Happiness," *Quantified Self*, April 15, 2013, http://quantifiedself.com/2013/04/konstantin-augemberg-on-tracking-happiness/.
7. "rTracker—Flexible Personal Data Tracking on the iPhone," Realidata, http://www.realidata.com/cgi-bin/rTracker/iPhone/rTracker-main.pl.
8. "Assessment Notes," Center of Inquiry at Wabash College, http://www.liberalarts.wabash.edu/ryff-scales/.
9. *Wikipedia*, s.v. "values scale," last modified March 29, 2015, https://en.wikipedia.org/wiki/Values_scale.
10. Ramirez, "Konstantin Augemberg on Tracking Happiness."
11. Carol D. Ryff, "Happiness Is Everything, or Is It? Explorations on the Meaning of Psychological Well-being," *Journal of Personality and Social Psychology* 57, no. 6 (1989): 1069–81, http://mina.education.ucsb.edu/janeconoley/ed197/documents/ryffHappinessiseverythingorisit.pdf.
12. Tricia A. Seifert, "The Ryff Scales of Psychological Well-being," 2005, http://www.liberalarts.wabash.edu/ryff-scales/.
13. Ibid.
14. Ibid.
15. Ryff, "Happiness Is Everything, or Is It?"
16. Shalom H. Schwartz, "Universals in the Content and Structure of Values: Theoretical Advances and Empirical Tests in 20 Countries," *Advances in Experimental Social Psychology* 25 (San Diego: Academic Press, 1992): 1–66.
17. "Schwartz," Pablo Gavilan, December 9, 2008, http://www.pablogavilan.com/tag/schwartz/.
18. H(app)athon Project, http://happathon.com/.
19. Peggy Kern, http://www.peggykern.org/.
20. Penn Positive Psychology Center, "Martin Seligman Bio," University of Pennsylvania, 2007, http://www.ppc.sas.upenn.edu/people/.
21. http://www.johnchavens.com/#!values/c19se.

22. Martin Seligman, *Flourish* (New York: Simon & Schuster, 2011).
23. Peggy Kern, ibid.
24. “Values Survey—Getting Insights from Your Data,” H(app)athon Project, April 15, 2014, http://happathon.com/values-survey-initial-insights-and-next-steps/.

Chapter Nine

1. Anthony Wing Kosner, “Tech 2015: Deep Learning and Machine Intelligence Will Eat the World,” *Forbes*, December 29, 2014, http://www.forbes.com/sites/anthonykosner/2014/12/29/tech-2015-deep-learning-and-machine-intelligence-will-eat-the-world/.
2. Andrew Y. Ng and Stuart Russell, “Algorithms for Inverse Reinforcement Learning,” Computer Science Division, UC Berkeley, 1999, http://ai.stanford.edu/~ang/papers/icml00-irl.pdf.
3. Ibid.
4. Stephen J. Dubner, “Time to Take Back the Toilet: A New Freakonomics Radio Podcast,” *Freakonomics*, December 18, 2014, http://freakonomics.com/2014/12/18/time-to-take-back-the-toilet-a-new-freakonomics-radio-podcast/.
5. Omohundro, “Autonomous Technology and the Greater Human Good.”
6. “The Universal Declaration of Human Rights,” United Nations, http://www.un.org/en/documents/udhr/. *Note*: Thanks to Steve Omohundro for alerting me to this document.
7. Ibid.
8. Ibid.
9. “The Universal Declaration of Human Rights: History of the Document,” United Nations, http://www.un.org/en/documents/udhr/history.shtml.
10. Future of Life Institute, “Research Priorities for Robust and Beneficial Artificial Intelligence.”
11. James Maynard, “Does Artificial Intelligence Really Threaten Mankind? Fears of Stephen Hawking, Elon Musk Stir Debate,” *Tech Times*, December 7, 2014, http://www.techtimes.com/articles/21672/20141207/artificial-intelligence-really-threaten-mankind-fears-stephen-hawking-elon-musk.htm.
12. Association for the Advancement of Artificial Intelligence, http://www.aaai.org/home.html.
13. 1st International Workshop on AI and Ethics, http://www.cse.unsw.edu.au/~tw/aiethics/AI-Ethics/Introduction.html.
14. Michael Anderson and Susan Leigh Anderson, “Toward Ensuring Ethical

Behavior from Autonomous Systems: A Case-Supported Principle-Based Paradigm," AAAI Fall Symposium Series (2014), http://www.aaai.org/ocs/index.php/FSS/FSS14/paper/view/9106/9131.

15. James Barrat, http://www.jamesbarrat.com/.
16. Omohundro, "Autonomous Technology and the Greater Human Good."
17. Stuart Russell and Peter Norvig, *Artificial Intelligence: A Modern Approach* (Harlow, UK: Pearson Education Limited, 2010).
18. Brockman, "The Myth of AI."
19. Eric Meyer, "Inadvertent Algorithmic Cruelty," *MeyerWeb.com*, December 24, 2014, http://meyerweb.com/eric/thoughts/2014/12/24/inadvertent-algorithmic-cruelty/.
20. Ibid.
21. AJung Moon's Profile, http://profile.amoon.ca/.
22. "ORI—About," Open Roboethics Initiative, http://www.openroboethics.org/about/.
23. "ORI—Technology," Open Roboethics Initiative, http://www.openroboethics.org/ori-code/.
24. Ergun Calisgan, AJung Moon, Camilla Bassani, Fausto Ferreira, Fiorella Operto, Gianmarco Veruggio, Elizabeth Croft, and H. F. Machiel Van der Loos, "Open Roboethics Pilot: Accelerating Policy Design, Implementation and Demonstration of Socially Acceptable Robot Behaviours," paper presented at We Robot: Getting Down to Business, University of Miami, 2013, http://conferences.law.stanford.edu/werobot/wp-content/uploads/sites/29/2013/04/Calisgan_Ergunv2.pdf.

Chapter Ten

1. NPR/TED Staff, "Can Hacking the Brain Make You Healthier?" *TED Radio Hour*, August 9, 2013, http://www.npr.org/2013/12/06/209618161/can-hacking-the-brain-make-you-healthier.
2. Amy Standen, "Hacking the Brain with Electricity: Don't Try This at Home," *Morning Edition*, May 19, 2014, http://www.npr.org/blogs/health/2014/05/19/312479753/hacking-the-brain-with-electricity-dont-try-this-at-home.
3. Foc.us, http://www.foc.us/.
4. Dr. Brent Williams, "The Foc.us tDCS Headset, Review Part 4, Electrode Placements," *SpeakWisdom*, August 31, 2013, https://speakwisdom.wordpress.com/2013/08/31/the-foc-us-tdcs-headset-review-part-4-electrode-placements/. *Note: I didn't actually purchase or try this headset but based my fictional narrative on information from the company's website and other reviews.*

5. Amit Amin, "The Science of Gratitude: More Benefits Than Expected; 26 Studies and Counting," *Happier Human*, http://happierhuman.com/the-science-of-gratitude/.
6. Gratitude 365, http://gratitude365app.com/.
7. Tomoko Tanaka, Yuji Takano, Satoshi Tanaka, Naoyuki Hironaka, Kazuto Kobayashi, Takashi Hanakawa, Katsumi Watanabe, and Manabu Honda, "Transcranial Direct-Current Stimulation Increases Extracellular Dopamine Levels in the Rat Striatum," *Frontiers in Systems Neuroscience* 7 (2013): 6, http://www.ncbi.nlm.nih.gov/pmc/articles/PMC3622879/.
8. Thai Nguyen, "Hacking into Your Happy Chemicals: Dopamine, Serotonin, Endorphins and Oxytocin," *Huffington Post*, October 20, 2014, http://www.huffingtonpost.com/thai-nguyen/hacking-into-your-happy-c_b_6007660.html. *Note: Again, I didn't actually purchase or try this headset but based my fictional narrative on information from the company's website and other reviews. I can't endorse or recommend this product but believe it could be used to test in the way I've indicated.*
9. "What Is Altruism?" *Greater Good*, http://greatergood.berkeley.edu/topic/altruism/definition.
10. Heroes, http://www.heroesu.com/#features.
11. Devpost, http://devpost.com/.
12. S. Lerner, C. M. A. Clark, and W. R. Needham, "A Simple Universal Basic Income Model for Canada Basic Income," *A Universal Basic Income*, http://www.basicincome.com/basic_model.htm.
13. Imogen Foulkes, "Swiss to Vote on Incomes for All—Working or Not," *BBC News*, December 18, 2013, http://www.bbc.com/news/business-25415501.
14. Noah Gordon, "The Conservative Case for a Guaranteed Basic Income," *The Atlantic*, August 6, 2014, http://www.theatlantic.com/politics/archive/2014/08/why-arent-reformicons-pushing-a-guaranteed-basic-income/375600/.
15. *Wonkblog, Washington Post*, http://www.washingtonpost.com/blogs/wonkblog/.
16. Is Giving Everyone a Check a Good Idea? *Wonkblog Debates*, April, 2014, http://www.washingtonpost.com/posttv/postlive/is-giving-everyone-a-check-a-good-idea/2013/12/05/6140ac7e-5dcd-11e3-95c2-13623eb2b0e1_video.html.
17. Negative Income Tax (NIT). https://en.wikipedia.org/wiki/Negative_income_tax.
18. Philip Harvey, "The Relative Cost of a Universal Basic Income and a Negative Income Tax," *Basic Income Studies* 1, no. 2 (2006), http://www.philipharvey.info/ubiandnit.pdf.
19. Karen DeWitt, "Service Rule Stirs Debate in Maryland," *New York Times*, July

29, 1992, http://www.nytimes.com/1992/07/29/education/service-rule-stirs-debate-in-maryland.html.

20. Stefan Klein, *Survival of the Nicest: How Altruism Made Us Human and Why It Pays to Get Along,* trans. David Dollenmayer (New York: The Experiment, 2014), 15.
21. Mihaly Csikszentmihalyi, *Flow: The Psychology of Optimal Experience* (New York: Harper & Row, 1990).
22. Ibid., 3.
23. *Wikipedia*, s.v. "gap analysis," last modified July 19, 2015, https://en.wikipedia.org/wiki/Gap_analysis.
24. "News & Events," Greater Good, http://bit.ly/1uJwBrC.
25. "Join the Greater Good Science Center," Greater Good, http://greatergood.berkeley.edu/membership.
26. Happify, http://www.happify.com.
27. Stan Neilson, *Robot Nation: Surviving the Greatest Socio-economic Upheaval of All Time* (Eridanus Press, 2011), http://www.amazon.com/Robot-Nation-Surviving-Greatest-Socio-economic/dp/0984150013.
28. Ibid.
29. Jeremy Bentham, "A Fragment on Government," *Constitution Society*, https://www.constitution.org/jb/frag_gov.htm.
30. *Wikipedia*, s.v. "normative ethics," last modified February 25, 2015, https://en.wikipedia.org/wiki/Normative_ethics.
31. Neilson, *Robot Nation*, 46.
32. Ford, *The Lights in the Tunnel.*
33. Peter Passell and Leonard Ross, "Daniel Moynihan and President-Elect Nixon: How Charity Didn't Begin at Home," *New York Times*, January 14, 1973, https://www.nytimes.com/books/98/10/04/specials/moynihan-income.html.
34. Marshall Brain, http://marshallbrain.com/.
35. "Robotic Nation," Marshall Brain, http://marshallbrain.com/robotic-nation.htm.
36. "Robotic Freedom," Marshall Brain, http://marshallbrain.com/robotic-freedom.htm.
37. "Manna," Marshall Brain, http://marshallbrain.com/manna1.htm.

Chapter Eleven

1. "Personal Robot," Kickstarter, https://www.kickstarter.com/projects/403524037/personal-robot.

2. Mike Jefferson, "Toto Smart Toilet Analyses Your Poo and Checks Your Health," *Gadget Lite*, August 2010, http://www.gadgetlite.com/2010/08/30/toto-smart-toilet-analyses-poo/.
3. Wit.ai, https://wit.ai/.
4. "Turn User Input into Action," Wit.ai, https://wit.ai/getting-started.
5. Wit.ai, https://wit.ai/.
6. "7 Foundation Principles," PbD, https://www.privacybydesign.ca/index.php/about-pbd/7-foundational-principles/.
7. "Inside ASIMO," ASIMO, http://asimo.honda.com/Inside-ASIMO/.
8. World Well-being Project, http://wwbp.org/.
9. "What Social Media Communicates About the Well-being of Society," *Knowledge@Wharton*, University of Pennsylvania, December 26, 2014, http://knowledge.wharton.upenn.edu/article/social-media-and-social-well-being/.
10. Sickweather, http://www.sickweather.com.
11. "Conference 2014," Gross National Happiness USA, http://www.gnhusa.org/conferences/conference-2014/.
12. Happiness Alliance, http://www.happycounts.org.
13. Happiness Alliance, http://survey.happycounts.org/survey/directToSurvey.
14. Laura Musikanski, "Happiness in Public Policy," *Journal of Social Change* 6, no. 1 (2014), http://scholarworks.waldenu.edu/jsc/vol6/iss1/5/.
15. John C. Havens, *Hacking H(app)iness: Why Your Personal Data Counts and How Tracking It Can Change the World* (New York: Tarcher, 2015), 222.
16. "World Happiness Report 2013," Issuu, http://issuu.com/rankings/docs/world_happiness_report_2013.
17. "Wealth vs. Well-being: How Do We Measure Prosperity?" Maryland, Smart, Green & Growing, http://www.dnr.maryland.gov/mdgpi/.
18. John C. Havens, "Abolish GDP in Favor of a Genuine Progress Indicator," *Guardian*, June 6, 2014, http://www.theguardian.com/sustainable-business/2014/jun/06/abolish-gdp-genuine-progress-indicator-gpi?CMP=twt_gu.
19. "Staff," Donella Meadows Institute, http://www.donellameadows.org/staff/.
20. Donella Meadows Institute, http://www.donellameadows.org/.
21. "Value of Housework," Maryland Department of Natural Resources, http://www.dnr.maryland.gov/mdgpi/17a.asp.
22. "American Time Use Survey," Bureau of Labor Statistics, http://www.bls.gov/tus/.
23. "Value of Housework."
24. E. F. Schumacher, *Small Is Beautiful: Economics as if People Mattered* (New York: Harper Perennial, 1989).
25. Ibid., 10.

26. Sankaran Krishna, "The Great Number Fetish," *Hindu*, January 26, 2013, http://www.thehindu.com/opinion/op-ed/the-great-number-fetish/article4345243.ece.
27. Adam Smith, *The Theory of Moral Sentiments* (Economic Classics, 2013).
28. "About the Library of Economics and Liberty," Library of Economics and Liberty, http://www.econlib.org/library/About.html#roberts.
29. Russ Roberts, *How Adam Smith Can Change Your Life: An Unexpected Guide to Human Nature and Happiness* (New York: Portfolio, 2014).

Chapter Twelve

1. Kevin Kelly, "The Three Breakthroughs That Have Finally Unleashed AI on The World," *Wired*, October 27, 2014, http://www.wired.com/2014/10/future-of-artificial-intelligence/.
2. Laurel D. Riek and Don Howard, "A Code of Ethics for the Human-Robot Interaction Profession," a paper presented at We Robot Conference, University of Miami, 2014, http://robots.law.miami.edu/2014/wp-content/uploads/2014/03/a-code-of-ethics-for-the-human-robot-interaction-profession-riek-howard.pdf.

INDEX

Note: The abbreviation "AI" refers to artificial intelligence.